Innovation, Networks, and Knowledge Spillovers

Manfred M. Fischer

Innovation, Networks, and Knowledge Spillovers

Selected Essays

With 20 Figures and 40 Tables

 Springer

Prof. Dr. Manfred M. Fischer
Vienna University of Economics and Business Administration
Institute for Economic Geography and GIScience
Nordbergstraße 15/4/A
1090 Vienna, Austria

ISBN-10   3-540-35980-X Springer Berlin Heidelberg New York
ISBN-13   978-3-540-35980-7 Springer Berlin Heidelberg New York

Cataloging-in-Publication Data
Library of Congress Control Number: 2006929605

Springer is a part of Springer Science+Business Media

springeronline.com

© Springer Berlin · Heidelberg 2006
Printed in Germany

Cover-Design: Erich Kirchner, Heidelberg

SPIN 11789390   88/3100-5 4 3 2 1 0 – Printed on acid-free paper

Preface

Our understanding of innovation processes has dramatically changed over the past decades. Interactive models of innovation, differing significantly from the linear approach, now emphasize both the importance of networking activities both within and across firms, and the centrality of knowledge spillovers which lie at the root of the formation of networks. The study of knowledge spillovers has become a major focus of research on innovative networks in recent years. Part of this research tradition can be traced back to the early work of Alfred Marshall on external economies through which 'industrial districts' – as he termed them – were integrated.

This book brings together a collection of articles and book chapters which present an overview and synthesis of current knowledge in the Economics of Innovation. It both reviews what is known and accepted as the best thinking on selected topics in the field and provides research findings that offer valuable insights into the nature and process of innovation, network formation and network activities, knowledge generation and spillovers from a regional perspective. By presenting the articles and book chapters as a whole, this collection is a novel combination. It is being published simultaneously with *Spatial Analysis and GeoComputation: Selected Essays*.

Innovation, Networks and Knowledge Spillovers is essentially a multi-product undertaking, in the sense that the various contributions are largely multi-authored publications. All these co-authors deserve the full credit for this volume, as they have been the scientific source of the research contributions included in the present volume.

I would also like to thank Gudrun Decker, Thomas Seyffertitz and Petra Staufer-Steinnocher for their capable assistance in co-ordinating the various stages of the preparation of the book.

Manfred M. Fischer   Vienna, May 2006

Contents

1 Introduction

Innovation is increasingly seen not only as a critical component in national and regional economic development but possibly *the* most important component. To a considerable extent, wealth creation is dependent on innovation. Accordingly many national and regional governments as well as academics have become closely focused on the factors and policies which may promote innovation (Storey 2004).

The concept of innovation has changed dramatically in recent years as the focus of attention has shifted from the single-act philosophy of innovation to the complex mechanisms that underline the production of new products and new production processes. At the same time, the earlier reference point, the linear research-to-marketing model of innovation, has been supplanted by interactive models of innovation which emphasise the central role of feedback effects between downstream [market related] and upstream [technology related] phases of innovation and the many interactions of innovative activities both within firms and in co-operative agreements among them (OECD 1992).

The interactive nature of the innovation process calls for organisational structures and mechanisms to ensure the appropriate interactions among the various institutions that make up spatial systems of innovation. This model underlines the importance of co-operation between firms and institutions, and, thus, the role played by networks involving different organisations. The growth of interfirm network activities represents a major change in the area of innovation.

Innovation is closely associated with knowledge. Indeed, innovation in a narrower sense may be defined as application of knowledge. Innovation does not only derive from the application of novel pieces of knowledge or a novel combination of existing pieces of knowledge, but knowledge can also be created during the process of innovation. It follows that learning – by individuals, teams and organisations – is highly important.

Specific forms of technological learning and knowledge creation, especially the tacit forms, are both localised and territorially specific. Firms that master non-codifiable knowledge are tied into various kinds of networks and organisations through localised input-output relations, especially knowledge spillovers and their untraded interdependencies (Storper 1997). Knowledge spillovers occur when knowledge created by a firm or organisation is not contained solely within that organisation, thereby creating value for other firms or organisations. The spillover beneficiary may use the new knowledge to copy or imitate the commercial products of the innovator, or may use the knowledge as an input to R&D leading to other new products or processes. The study of knowledge spillovers has become a major focus of research on networks and clusters in recent years.

The contributions to this volume collection have been carefully chosen in order to give readers an overview and synthesis of key themes in the areas of Innovation, Networks and Knowledge Spillovers, from a regional perspective. We have chosen articles and book chapters which we feel should be made accessible not only to specialists but to a wider audience as well. Thus, we hope that the volume will be useful both for undergraduate ad graduate students, for senior researchers, as well as for persons outside academia such as public policy makers.

By bringing together this specific selection of articles and book chapters and by presenting them as a whole, this collection is a novel combination. In trying to select among many interesting articles, we decided that there should be both, overviews and state-of-the-art pieces on the subject as a whole, as well as certain pieces which focus on particular issues of interest to readers. Some articles selected are not well known but develop an issue in an interesting way, or add a new twist on existing controversies.

The purpose of this introduction is to serve as a brief guide to the material presented in this collection. The introduction will, thus, set the scene by organising the themes, revealing links and also by helping to locate the contributions within the book. The book is organised in three parts. PART I demonstrates that the processes of innovation and technological change are spatially differentiated, regionally within countries and internationally between countries. PART II broadens, both conceptually and empirically, our understanding of the innovation process and the process of network formation, by examining in particular the increasing importance of knowledge creation and diffusion in the new economy and how this is changing the nature of firms in crucial ways. Particular focus is laid on identifying the growing pressures for firms to develop more inter- and intrafirm networks and on providing lucid illustrations of these different kinds of networks. PART III focuses on knowledge creation and spillovers, and presents important, enlightening conceptual and empirical work on the systems of innovation approach to innovation analysis, the regional knowledge production function approach and the case-control matching approach to study the spatial dimension of knowledge spillovers.

PART I   Innovation and Technological Change

PART I of the present volume is composed of three contributions:

- Innovation and Technological Change: An Austrian-British Comparison (Chapter 2),
- Technology, Organisation and Export-Driven Research and Development in Austria's Electronics Industry (Chapter 3) and
- Information-Processing, Technological Progress and Retail Market Dynamics (Chapter 4).

These three contributions represent very different styles of research. While the first two chapters are primarily directed towards empirical issues, no attempt is made in the final chapter to address empirical issues. This chapter has been chosen to represent the mathematical richness when analysing technological change in general and specific implications in particular within a more formal modelling framework. The focus of the chapter is on the implication of the technological progress in information processing on the size of retail markets.

Although there is a larger body of empirical evidence available that demonstrates the nature of spatial variations in innovation and the adoption of new technologies at the time of writing this chapter, only few studies had been conducted in such a way as to enable direct comparisons between different countries, either to establish international differences in innovative performance or to identify differences in regional patterns in different institutional contexts. **Chapter 2**, written with Neil Alderman [Centre for Urban and Regional Development Studies, University of Newcastle upon Tyne], makes use of the results of survey of comparable industries in Great Britain and Austria to address this issue, with particular attention to some of the inherent difficulties in undertaking such comparisons. By using a mixture of single cross-tabulations and multivariate logit models, cross-country differences are identified in the adoption of a number of new process technologies based upon microelectronics in the spheres of manufacturing production, design, and co-ordination. It is suggested that not only does Austria lag Great Britain in the introduction of new technology, but that variations between similar types of regions are also more pronounced and entrenched in Austria.

Chapter 3, written jointly with Luis Suarez-Villa [School of Social Ecology, University of California, Irvine], moves to consider the relationship between R&D, territorial location and the most important organisational characteristics in Austria's electronics industries. An assessment of operational motivations, based on establishment-level survey data, is followed by a factor analysis that reveals the main organisational dimensions. Statistical analyses of the association between R&D intensity, localities and the organisational factors are then expanded to consider subcontracting and just-in-time production methods. Two-way sub-contracting, whereby firms both subcontract part of their production out and are in turn contracted by others, is found to be prevalent. Such arrangements are thought to help firms specialise and avoid implementing costly production techniques whilst helping save capital and resources that can be reinvested in R&D. These analyses provide important insights on the association between R&D and subcontracting, just-in-time production and on the advantages of skilled production labour and plant size for research-intensive manufacturing establishments.

The final chapter in PART I [**Chapter 4**], written jointly with Jacek Cukrowski [Center for Social and Economic Research, Warsaw], analyses the implication of technological progress in information processing on the size of retail markets. The analysis – restricted to a single commodity market with uncertain demand – shows that the ability of firms to process information and predict demand may affect the characteristics of retail markets. The results, moreover, indicate that risk-averse firms always devote resources to demand forecasting, producers are better off trading with retailers than with final consumers, and the volume of output supplied through retail markets is greater than it would be if producers traded directly with consumers.

PART II   Innovation and Network Activities

One of the most fundamental characteristics of the emergence of the knowledge based economy (see Ács et al. 2002 for a discussion) is the growing extent to which actors need to co-operate more actively and more purposefully with each other in order to cope with increasing market pressures stemming from global competition and rapid technological change (Suarez-Villa 2002). This is evidenced by both the growth of interfirm networks and the closer integration of research, development, production, and marketing within the company. Firms co-operate in order to gain rapid access to new technologies or markets, to benefit from economies of scale in joint R&D and production, to tap into external sources of know-how, and to share risks. The need for co-operation stems from the specific nature of knowledge – as distinct from information – in particular, its tacit and specific features. Three contributions have been chosen for PART II with a focus on *Innovation and Network Activities*. These are as follows:

- The New Economy and Networking (Chapter 5),
- The Innovation Process and Networking Activities of Manufacturing Firms (Chapter 6), and
- Knowledge Interactions between Universities and Industry in Austria: Sectoral Patterns and Determinants (Chapter 7).

The first chapter in PART II, **Chapter 5**, is a review paper that attempts to connect the knowledge based economy with the formation of networks by examining the increasing importance of knowledge creation and diffusion in the new economy and how this is changing the nature of the firms in crucial ways. The paper, in particular, identifies the growing pressures for firms to develop more intra- and interfirm networks, provides lucid illustrations of these different kinds of networks and describes the emergence of the network enterprise of corporate organisation.

Chapter 6 continues to contribute to our understanding of both the innovation process and the process of network formation by emphasising that the interactive nature of the innovation process has broken down the distinction between innovation and diffusion so that the creation of knowledge and its assimilation via networks are part of a single process. The discussion is enriched with empirical evidence illustrating the importance and diversity of interfirm network activities of manufacturing firms in the metropolitan region of Vienna. The picture which emerges from the evidence of the study is that there exists a maze of different networks. They range from highly formalised to informal network relations, from highly specialised and rather narrow networks to much wider and looser networks such as, for example, technical alliances involving firms as corporate entities, from networks focusing on the pre-competitive stage of the innovation process to those involving the competitive stage. Co-operation in the early stages of the innovation process is found to be generally more common than in the competitive stage. External information tends to be particularly relevant in the pre-competitive stage, when perception of problems and evaluation of technological possibilities take place. Customer and user-producer (i.e., manufacturing and producer service supplier) relationships are frequently used, horizontal forms of co-operation (such as producer network linkages) and collaboration with public research organisations to a lesser extent. Producer network linkages are typically found among smaller enterprises that may occasionally also be competitors.

Chapter 7, jointly written with D. Schartinger, C. Rammer and J. Fröhlich [Austrian Research Centers Seibersdorf], moves attention to R&D related co-operation and communication between universities and industry. The relationship between industry and the academic world is a complex and heterogeneous phenomenon, and an important topic of the recent debate on innovation systems. Chapter 7 is an empirical study based on direct measurement of the extent of collaboration between firms and universities in the Austrian system of innovation. Unlike previous studies this chapter uses information on both the importance of different channels of knowledge exchange [from the universities' view] and, most importantly, the number of R&D projects conducted jointly with firms in the years before the survey. The study provides valuable insight into several dimensions of knowledge flows between universities and firms that are not typically explored in research on this topic. The patterns of interaction identified between 46 different fields of science and 49 economic sectors corresponding to the NACE classification at the 2-digit level represent an important and interesting outcome of the analysis. The data show, for example, that some non-technical disciplines have considerable interactions with firms, sometimes with unexpected sectors. Left censored Tobit models are used to evaluate the effect of sector specific and science field specific characteristics upon the probability of knowledge interactions, disaggregated by type of interaction.

PART III   Knowledge Creation, Diffusion and Spillovers

Knowledge creation, knowledge diffusion and spillovers are central to systems of innovation. A system of innovation may be thought as a set of actors such as firms, other organisations [such as universities, for example], and institutions that interact in the generation, diffusion and use of knowledge in the production process. Institutions may be viewed as sets of common habits, routines, established practices, rules or laws that regulate the relations and interactions between individuals within as well as between and outside the organisations.

The systems of innovation approach that has emerged during the last decade or so (see Freeman 1987, Lundvall 1992, Nelson 1993, Edquist 1997) has diffused fast in the academic world as well as in the realms of public innovation policy making. The approach is at the centre of modern thinking about innovation and the relations of innovation to economic growth and competitiveness. The approach is not considered to be a formal and established theory, but its development has been influenced by different theories of innovation such as interactive learning theories and evolutionary theories, and by the idea of a 'national system of innovation'.

The focus of PART III is on *knowledge creation, diffusion and spillovers*. The following four papers have been chosen for the final part of the book:

- Innovation, Knowledge Creation and Systems of Innovation (Chapter 8),
- The Role of Space in the Creation of Knowledge in Austria. An Exploratory Spatial Data Analysis (Chapter 9),
- Spatial Knowledge Spillovers and University Research: Evidence from Austria (Chapter 10) and
- Patents, Patent Citations and the Geography of Knowledge Spillovers in Europe (Chapter 11).

Chapter 8 critically reviews the systems of innovation approach as a framework for regional innovation research and presents an effort to both develop some missing links and decrease the conceptual noise often present in the discussions on national innovation systems. The chapter specifies elements and relations that seem to be essential to the conceptual core of the framework and argues that there is no a priori reason to emphasise the national over the subnational [regional] scale as an appropriate mode for analysis, irrespective of time and place. Localised input-output relations between the actors of the system, knowledge spillovers and their untraded interdependencies lie at the centre of the argument.

Knowledge spillovers occur because knowledge created by a firm or other organisation is not normally contained solely within that organisation, but is also exploited by other firms. Three vehicles of such spillovers may be distinguished: *first*, the scientific sector with its general scientific and technological knowledge pool; *second*, the firm specific knowledge pool; and, *third*, the business-business and university-industry relations that make them possible.

The relationship between knowledge spillovers and space is extremely complex and, at the current state of research, only partially understood. This is partly due to the fact that knowledge spillovers are notoriously difficult to measure. **Chapter 9**, written jointly with J. Fröhlich and H. Gassler [Austrian Research Centers Seibersdorf] and A. Varga [now at the University of Pecs], makes a modest attempt to shed some light on the role of space in the creation of technological knowledge in Austria. The study is exploratory rather than explanatory in nature and based on descriptive techniques such as Moran's I test for spatial autocorrelation and the Moran scatterplot. Clusters of the knowledge output [measured in terms of patent counts] are compared with spatial concentration patterns of two input measures of knowledge production: private R&D and academic research. In

addition, employment in manufacturing is utilised to capture agglomeration economies. The analysis is based on data aggregated for two-digit SIC industries and at
the level of Austrian political districts. It explores the extent to which knowledge
spillovers are mediated by spatial proximity in Austria. A time-space comparison
makes it possible to study whether divergence or convergence processes in knowledge creation have occurred in the past two decades.

Most studies identifying the spatial extent of knowledge spillovers are based on
the Griliches-Jaffe knowledge production function model (see Griliches 1979,
Jaffe 1989) to measure knowledge spillovers, indirectly via effects on the output
of the knowledge production function. This type of research, however, is not
without problems. The problems center around the question of whether the spatial
units of observation are appropriately chosen, whether and how spatial effects are
taken into account, how the output of the knowledge production process is
measured, whether available measures actually capture the contribution of R&D
spilled-over, how the spillover pools are constructed, and R&D capital defined
and depreciated. Despite these difficulties, there has been a significant number of
reasonably well done studies, all pointing in the direction that knowledge
spillovers tend to be geographically bounded within the region of knowledge
production.

Chapter 10, written with A. Varga [now at University of Pecs], provides an
example of this research direction. The study provides evidence on the issue of
geographically mediated knowledge spillovers from university research activities
in high-tech industries in Austria. It is assumed that knowledge production in the
high-technology sectors essentially depends on two major sources of knowledge:
the university research that represents the potential pool of knowledge spillovers
and R&D performed by the high-tech sectors themselves. Using district-level data
and refining the classical regional knowledge production function, and employing
spatial econometric tools, evidence is found that transcend the geographic scale of
the political district in Austria. It is shown that geographic boundedness of the
spillovers is linked to a decay effect.

Localisation of knowledge spillovers is implicit in most theories of new
economic growth, but rarely studied empirically. One notable exception is the
pioneering analysis of Jaffe et al. (1993) on patent citations. The final contribution
to PART III, **Chapter 11**, written with T. Scherngell and E. Jansenberger
[Institute for Economic Geography and GIScience, Vienna University of
Economics and Business Administration], is similar in spirit and directs attention
to knowledge spillovers between high-technology firms in Europe, as captured by
patent citations. The study adopts the case-control matching approach to test the
extent of localisation of knowledge spillovers and finds strong evidence of
geographic localisation at two spatial levels [country, region] even after controlling for the tendency of inventive activities in the high-technology sector to be
geographically clustered. The findings not only indicate that knowledge spill-over
localisation exists in the aggregate, but that there are also variations of localisation
by region.

Conclusions. This collection of chapters we have selected covers the most
recent past of research. They draw upon and reveal research extending over a

period of about fifteen years. The issues addressed are far more numerous, complex, fascinating and intriguing than one might initially imagine. The studies presented are hoped to set the stage for novel, promising and – perhaps – more sophisticated applied research in this fascinating scientific domain that involves the most vibrant issues and debates of our time.

References

Ács Z. (ed.) (2000): *Regional Innovation, Knowledge and Global Change*, Pinter, London

Ács Z., Groot H.L.F. de and Nijkamp P. (eds.) (2002): *The Emergence of the Knowledge Economy. A Regional Perspective*, Springer, Heidelberg, Berlin, New York

Barney J.B. (1986): Strategic factor markets: Expectations, luck, and business strategy, *Management Science* 32, 1231-1241

Bertuglia C.S., Fischer M.M. and Preto G. (eds.) (1995): *Technological Change, Economic Development and Space,* Springer, Berlin, Heidelberg, New York

Castells M. (1996): *The Rise of the Network Society*, Blackwell, Oxford [UK], Malden [MA]

Chesnais F. (1988): Technical cooperation agreements between firms, *STI Review* 4, 51-120

DeBresson C. and Amesse F. (1991): Networks of innovators: A review and introduction to the issue, *Research Policy* 20, 363-379

Dosi G., Freeman C., Nelson R.R., Silverberg G. and Soete L. (eds.) (1988): *Technical Change and Economic Theory,* Pinter, London

Edquist C. (ed.) (1997): *Systems of Innovation: Technologies, Institutions and Organisations*, Pinter, London

Fischer M.M. and Fröhlich J. (eds.) (2001): *Knowledge, Complexity and Innovation Systems*, Springer, Heidelberg, Berlin, New York

Fischer M.M., Revilla-Diez J. and Snickars F. (2001): *Metropolitan Innovation Systems*, Springer, Heidelberg, Berlin, New York

Freeman C. (1994): The economics of technological change, *Cambridge Journal of Economics* 18, 463-514

Freeman C. (1987): *Technology and Economic Performance: Lessons from Japan*, Pinter, London

Griliches Z. (1979): Issues in assessing the contribution of research and development to productivity growth, *Bell Journal of Economics* 10, 92-116

Håkansson H. (1987): *Industrial Technological Development: A Network Approach,* Croom Helm, London

Jaffe A.B. (1989): Real effects of academic research, *American Economic Review* 79, 957-970

Jaffe A.B., Trajtenberg M. and Henderson R. (1993): Geographic localization of knowledge spillovers as evidenced by patent citations, *Quarterly Journal of Economics* 108 (3), 577-598

Lundvall B.-Å. (ed.) (1992): *National Systems of Innovation: Towards a Theory of Innovation and Interactive Learning*, Pinter, London

Malecki E.J. (1997): *Technology & Economic Development. The Dynamics of Local, Regional and National Competitiveness*, Longman, Harlow, Essex

Mansfield E. (1995): Academic research underlying industrial innovations: Sources, characteristics and financing, *The Review of Economics and Statistics* 77, 55-65

Nelson R.R. (ed.) (1993): *National Innovation Systems: A Comparative Analysis*, Oxford University Press, Oxford [UK]

Nelson R.R. and Winter S.G. (1982): *An Evolutionary Theory of Economic Change*, Belknap Press of Harvard University Press, Cambridge [MA]

Nijkamp P. and Poot J. (1998): Spatial perspectives on new theories of economic growth, *The Annals of Regional Science* 32 (1), 7-37

OECD (1992): *Technology and the Economy*, Organisation for Economic Co-operation and Development, Paris

Storey J. (ed.) (2004): *The Management of Innovation. Volume 1*, Edward Elgar, Cheltenham [UK]

Storper M. (1997): *The Regional World. Territorial Development in a Global World*, The Guilford Press, New York, London

Suarez-Villa L. (2002): Networked alliances and innovation. In: Ács Z. and Groot H.L.F. de and Nijkamp P. (eds.) (2002): *The Emergence of the Knowledge Economy. A Regional Perspective*, Springer, Heidelberg, Berlin, New York, pp. 65-80

Suarez-Villa L. (1989): *The Evolution of Regional Economies. Entrepreneurship and Macroeconomic Change*, Praeger, New York, Westport, London

Teece D.J. (1988): Technological change and the nature of the firm. In: Dosi G., Freeman C., Nelson R.R., Silverberg G. and Soete L. (eds.) (1988) *Technical Change and Economic Theory*, Pinter, London, pp. 256-281

Wernerfelt B. (1984): A resource-based view of the firm, *Strategic Management Journal* 5, 171-180

Part I

Innovation and Technological Change

2 Innovation and Technological Change: An Austrian-British Comparison

with *N. Alderman*

Despite a growing body of empirical evidence that demonstrates the nature of spatial variations in innovation and the adoption of new technologies, at the time this chapter was written few studies had been conducted in such a way as to enable direct comparisons between different countries, either to establish international differences in innovative performance or to identify differences in regional patterns in different national contexts, particularly between EC and non-EC countries within Europe. In this paper the results of recent surveys of comparable industries in Great Britain and Austria are used to begin to address this issue, with particular attention to some of the inherent difficulties in undertaking such comparisons. By using a mixture of simple cross-tabulations and multivariate logit models, differences between the two countries in the adoption of a number of new process technologies based upon microelectronics in the spheres of manufacturing production, design, and co-ordination are identified. It is suggested that, not only does Austria lag Great Britain in the introduction of new technology, but that variations between similar types of region are more pronounced and entrenched in Austria at that time.

1 Introduction

One of the key features of the current wave of technological change taking place within manufacturing is the adoption of microelectronics-based technologies in traditional manufacturing processes. Such technologies may be roughly defined as comprising all those new technologies which use microprocessors or their electronic equivalents (such as custom or semicustom integrated circuits) either in the form of single integrated circuit devices or in small groups of linked devices. The micro-electronic revolution is not only creating new goods and services, but also altering how they are produced. In manufacturing, microprocessors have gradually penetrated into all aspects of the production process. Applications cover the use of microelectronics-based equipment in design, fabrication, assembly, handling, quality control and testing, or other operations on site necessary to make a product ready for sale. Typical process and production applications include the use of computer-aided design (CAD) equipment, computer-aided manufacturing (CAM) systems, including inter alia computerised numerically controlled (CNC) machine tools, robots, and flexible manufacturing systems (FMS). In contrast to special-purpose automated machines these programmable automation technologies tend to increase flexibility and efficiency (in terms of both the range of products and the volume of a specific product) as well as to increase productivity and control over the manufacturing process (see Fischer 1990).

Over the past ten years a considerable amount of empirical evidence of one sort or another has been amassed that demonstrates that these processes of innovation and technological change are spatially differentiated, both regionally within nations and internationally between nations (for example, Brugger and Stuckey 1987; Jacobsson 1985; Kleine 1982; Nabseth and Ray 1974; Rees et al. 1984; Thwaites et al. 1982; Tödtling 1990). Few studies, however, have been conducted in such a way as to enable direct comparisons between countries, either to establish international differences in the innovative performance of particular industries or to identify differences in regional patterns within different national contexts.

Reliable cross-national comparisons will become an increasingly pressing need as the issue of European integration rises higher on the political and economic agenda. The implementation of the Single European Act in 1993 and the changing climate of East-West relations will add urgency to this issue (Commission of the European Communities 1991; Quévit 1991).

Inconsistencies between national studies in terms of survey design – sectoral composition, choice of innovations, categorisations of variables, and so forth – mean that it is frequently impossible to conclude whether differences (or conversely similarities) between national experiences can be attributed to fundamentally different levels of industrial performance, or different economic, political, and cultural regimes, or whether they are simply the product of different sample designs.

In this paper, evidence from recent surveys of comparable industries in Austria (Fischer and Menschik 1991) and Britain (Alderman et al. 1988; Thwaites et al. 1982) is used to investigate the comparative innovative performance of the two countries in terms of the adoption of some of the key computer-based technologies referred to above. By controlling for variables such as industrial sector, the comparative performance of manufacturing in similar types of region (the core metropolitan region and its immediate hinterland, a traditional iron-based industrial region, and a peripheral region) is identified.

By using appropriate multivariate analyses, the importance of the commonly identified indicators of innovation propensity is tested and the difference between Austrian and British manufacturing establishments identified. The prospects for the different types of regions in the two national settings in terms of the adoption of the components of computerised manufacturing systems are discussed.

2 Methodology

Previous international comparisons have been concerned with the differences between leading industrial nations, such as the USA, West Germany, and the

United Kingdom (for example, Gibbs and Thwaites 1985) or between the core nations of the European Economic Community (EC) (Northcott et al. 1985). Comparisons between other nations within Europe, particularly between EC and non-EC countries, are largely missing. We attempt to fill a gap in this aspect of international comparison by comparing adoption experiences in Great Britain, an EC country, with those of Austria, a non-EC country. It raises the question of whether the types of spatial disparities in technological development observed within core nations are replicated in more peripheral nations and whether they occur with similar levels of intensity.

The research in Austria was undertaken at the Department of Economic Geography of the Vienna University of Economics and Business Administration, funded by the Jubiläumsfonds provided by the Austrian National Bank. Data on the spatial pattern of the adoption of specific techniques within a limited number of manufacturing industries were obtained through interview surveys of senior executives of manufacturing establishments and enterprises. The survey was designed to explore in greater depth the characteristics of adopting and non-adopting establishments, including their approach to technology and investment generally, as well as their reasons for adoption or non-adoption of the specified techniques. In the interviews, we also investigated the sources of information used to evaluate technological change, changes in labour requirements related to technology, and the use of government aid in the adoption process. By use of a questionnaire, we also obtained information concerning the ownership of the establishment, its employment size, the extent of R&D activity, etc.

The data were obtained from establishments in the Austrian metalworking and machinery, electrotechnical and electronic products, textiles and clothing industries. Owing to time and resource constraints, the interviews were limited to four Austrian regions only: the core metropolitan area of Vienna, its immediate hinterland, a traditional iron-based industrial region (Upper Styria) and a peripheral region (Wald-Weinviertel) (see Figure 1) which represent a variety of historic and current economic trends and conditions within the Austrian economy. A total of 185 interviews, each lasting about two hours, were conducted between November 1987 and February 1988 with senior industrialists who were manufacturing in the selected regions (see Fischer and Menschik 1991).

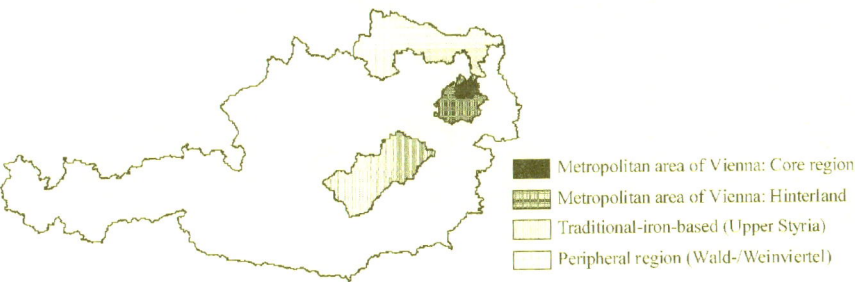

Metropolitan area of Vienna: Core region
Metropolitan area of Vienna: Hinterland
Traditional-iron-based (Upper Styria)
Peripheral region (Wald-/Weinviertel)

Figure 1   Study areas in Austria

The research in Britain was undertaken by the Centre for Urban and Regional Development Studies at the University of Newcastle upon Tyne and forms part of a long-running research programme into the spatial dimension to technological change. The data presented here were collected in two surveys of establishments in a range of metalworking industries within Great Britain (see Table 1).

The first survey was undertaken in 1981 and formed part of a project funded by the UK Department of Trade and Industry and by the Regional Directorate of the EC (Thwaites et al. 1982). Data collection was primarily by means of a postal questionnaire, but this was supplemented by interviews with executives in 130 establishments in four regions (the South East, West Midlands, the North, and Scotland). The second survey was a follow-up to the first (that is, no new establishments were surveyed) and took place by means of a postal questionnaire in 1986-87, which was followed up by telephone during 1987 and 1988. This latter survey concentrated on identifying adopters of new technologies for the purposes of testing forecasts of technology diffusion at the regional level (see Alderman et al. 1988). As such, this survey was not very detailed, but a final response rate of over 95% of surviving establishments was achieved. In the analysis that follows only those establishments surviving through to 1986 are included.

Table 1   Standard industrial classifications (SICs) for the Austria-Britain comparison

Austria	Great Britain (1968 SIC)
SIC Description	SIC Description
Metalworking and machinery	
51 Manufacturing of iron and nonferrous metals	331 Agricultural machinery
52 Machining of metals, steel-girder and light-metal construction	332 Metalworking machine tools
53 Manufacturing of hardware	333 Pumps, valves, and compressors
54/55 Manufacturing of machines (excluding electric machines)	336 Contractors' plant and machinery
58 Manufacturing of means of transportation	337 Mechanical handling equipment
	339 General mechanical engineering
	341 Industrial plant and structural steelwork
	390 Engineers' small tools and gauges
Electrotechnical and electronic products	
56/57 Manufacturing of electrical installations	361 Electrical machinery

In recent years manufacturing industry has experienced rapid technological changes which have been focused upon process innovations utilising the advances in microelectronics. In the two studies, we examined the spatial diffusion of selected process innovations which are of particular relevance to the metalworking and machinery industry as well as to the electrotechnical and electronic products industry. The selection of the industries and production techniques for the comparison was an interactive process. The techniques were selected on the basis that they introduced fundamental rather than minor incremental change, were economically significant, and had a comparatively recent diffusion pattern. The selected techniques providing the foci of the comparison are:

- numerically controlled (NC) and CNC machine tools;
- computers for design (CAD) and computer-aided engineering (CAE);
- computers for manufacturing operations (CAM, CAD-CAM) and microprocessors;
- computers for commercial use.

There are some minor differences in technology definition that should be noted. The British follow-up survey was concerned specifically with CAD and drafting systems rather than with computers in the design sphere more generally. Nevertheless, these types of CAD system are the most prevalent and rapidly diffusing applications at the present time.

In the manufacturing sphere the British follow-up survey dropped the broad definition of computers for manufacturing operations on the grounds that this was too vague a definition, concentrating instead on the adoption of microprocessors in the manufacturing process. In other respects the Austrian and British surveys are identical as far as technology definition goes.

2.1 Regional Comparison

The British survey was a national one, in contrast to the Austrian study which was limited to the regions outlined above. It was therefore necessary to identify suitable areas within Great Britain that would provide a reasonable match for comparative purposes. In the event, the choices rested largely on the pragmatic considerations of which areal units were available and the numbers of observations involved (see Figure 2).

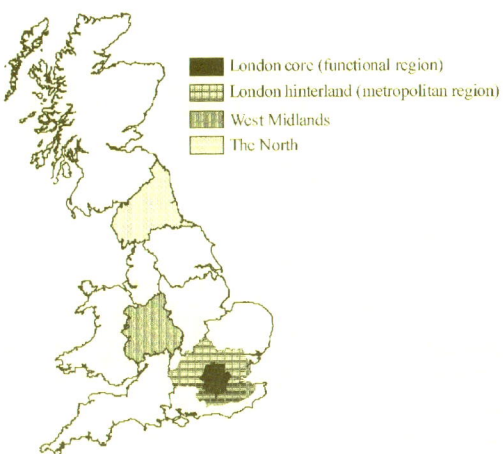

London core (functional region)
London hinterland (metropolitan region)
West Midlands
The North

Figure 2   Study areas in Great Britain

The core metropolitan region of Vienna was matched against the London functional region as defined by Coombes et al. (1982), and the hinterland was matched against the rest of the London metropolitan region on the basis of the same regionalisation. As such, the London regions are rather larger than those for Vienna, but this is unavoidable as the equivalent administrative and built-up areas are also considerably larger. These size differences in population terms are illustrated in Table 2.

For the remaining two areas, the West Midlands standard region was matched against Upper Styria as it is the home of the iron-based industries in Britain, and the northern standard region was chosen as a representative peripheral region. The major difference in the case of the North is that most industrial activity is centred on the major conurbations of Tyneside and Teesside, which have no equivalent in Wald-Weinviertel.

Table 2 demonstrates that the British standard regions are rather larger than the Austrian regions. Nevertheless, there are some similarities in that the traditional iron-based regions both have high levels of manufacturing employment (although the same is true of the Vienna hinterland in contrast to that for London). Austrian unemployment in 1981 was much lower than in Britain, although it worsened during the 1980s as in other advanced economies. Moreover, the peripheral region chosen was not experiencing the highest levels of unemployment in Austria the way it was in Britain and the radically different economic problems in these regions, the former figure reflecting a predominantly rural economy, the latter reflecting a predominantly urban problem.

Table 2   Regional comparisons

Area	Population in 1981	Manufacturing employment 1981 (%)	Unemployment rate in 1981
London core	7,665,455	19.6	8.3
Vienna core	1,532,344	27.1	2.1
London hinterland	4,494,072	28.6	5.8
Vienna hinterland	285,936	35.3	2.8
British iron-based region	5,112,349	39.2	11.7
Austrian iron-based region	280,067	34.5	2.9
British peripheral region	3,090,404	30.5	13.3
Austrian peripheral region	278,067	18.4	3.3

2.2 Comparison of the Samples

As a result of the different spatial sampling schemes used it is not surprising to find the composition of the two samples to be different. Table 3 shows that in the Austrian case the sample is dominated by the metropolitan area, whereas in the British case the iron-based and peripheral regions take the 'lion's share'. The other major distinction, of course, is that the Austrian survey was not large, but extremely detailed, whereas the British survey was large, but limited in terms of the information collected, and this inevitably affected subsequent analysis.

Table 3 Composition of Austrian and British samples (percent of establishments) by regional type. Sources: national surveys (Austria, November 1987; $N = 136$. Great Britain, 1986; $N = 262$) (see Fischer and Menschik 1991; Alderman et al. 1988)

Area	*Percent of establishments*	
	Austria	*Great Britain*
Metropolitan area	63.2	37.4
Core region	36.8	18.7
Hinterland	26.4	18.7
Traditional iron-based industrial region	19.9	34.4
Peripheral region	16.9	28.2

Although an attempt has been made in the two studies to control for sectoral differences (differences in national industrial classifications inevitably cause problems; for example, see Gibbs and Thwaites 1985) by focusing on industries engaged in similar activities (metalworking, electrical equipment, machinery) which may therefore be expected to have broadly similar opportunities and requirements for the adoption of new technology, other factors related to the structure of these industries in the two countries could be influential. One of these factors is the presence in Austria of a strong nationalised sector (the so-called Austrian Industries), which in the industries surveyed accounts for over 10% of employment. Table 4 indicates, however, that in the Austrian case there are rather more independent establishments. Comparing 1981 with 1986 information, the British sample shows a decline in the proportion of independent establishments during the 1980s (and this despite an increasing number of management 'buyouts'). The proportion of branch plants in the British sample was rather higher on the basis of 1981 information.

Table 4 Corporate status composition of Austrian and British samples (sources: see Table 3)

Plant description	*Percent of establishments*		
	Austria	*Great Britain*	
		1986	1981
Single-plant enterprise	50.7	40.1	45.2
Multiplant organisation	49.3	59.9	54.6
Head office	15.4	nd	nd
Divisional headquarter	9.6	nd	24.0
Regional headquarter	10.3	nd	nd
Branch	14.0	nd	30.6

nd no data available

The age structure of the two samples also shows differences, primarily because no establishments starting up after 1981 were identified in the British survey. Table 5 shows how in Austria the age distribution is skewed towards very young establishments, whereas in Britain the skew is towards older establishments.

Table 5   Variation in establishment age by country (sources: see Table 3)

Period of establishment	Percent of establishments	
	Austria	Great Britain
Pre-1950	21.5	41.3
1950-59	23.0	16.2
1960-69	14.8	25.1
1970-79	8.1 }	17.4
1980 and later	32.6 }	

The most important distinctions are likely to be in terms of the size distributions, not only because of the theoretical importance of size in technology adoption (for instance, Davies 1979; Freeman 1974), but also because of its observed empirical importance (Alderman et al. 1988; Northcott and Rogers 1984; Rees et al. 1984; Thwaites et al. 1982). Surprisingly, perhaps, the sample-size distributions appear to be similar, but there are more very small establishments in Austria, and more in the 100-499 (number of employees) category within the British sample (Table 6). On this basis alone we should anticipate higher adoption levels in the British context. It is to the national and regional differences in levels of new-technology adoption that we now turn.

Table 6   Differential employment size structures (sources: see Table 3)

Country	Percent of establishments by employment size[a]			
	1-49	50-99	100-499	≥ 500
Austria	44.9	18.4	25.0	11.8
Great Britain	38.0	20.2	31.4	10.5

[a] number of employees

3   The Adoption of New Technology

Variations in technological change between countries and regions can be anticipated simply as a result of the differing nature of the enterprises and establishments operating therein (Thwaites 1978). In this section evidence is provided of the extent of adoption of the selected technologies and these are related to the characteristics of the establishments in each country.

In crude terms, we demonstrate in Table 7 that there are substantial differences in adoption levels between the two countries and between the regions within them. In general, with the exception of computers for commercial uses, adoption levels are higher in the British case, although to some extent this is expected, because of the differences in size distribution. However, even in 1981, levels of NC adoption amongst surviving British establishments were considerably higher than they were in Austria in 1987. CAD adoption similarly appears to be further advanced in Britain than in Austria.

Table 7   Adoption of new technology (percent of establishments) by regional type (sources: see Table 3)

Technology	Metropolitan core		Metropolitan hinterland		Traditional iron-based industrial		Peripheral	
	Austria	Britain	Austria	Britain	Austria	Britain	Austria	Britain
NC machines[a]	14.0	32.7	16.7	38.8	3.7	34.4	17.4	27.0
CNC machines	20.0	49.0	22.2	59.2	40.7	60.0	4.3	44.6
CAD, CAE	14.0	20.4	8.3	38.8	11.1	28.9	8.7	18.1
CAM, CAD–CAM	12.0	nd	5.6	nd	18.5	nd	8.7	nd
Microprocessors	nd	24.5	nd	40.8	nd	41.4	nd	24.3
Computers for commercial use	88.0	77.6	75.0	81.6	74.0	81.1	60.9	75.7

[a] 1981 rate for Britain; nd no data available; NC numerically controlled; CNC computer numerically controlled; CAD computer-aided design; CAE computer-aided engineering; CAM computer-aided manufacturing

Regional discrepancies appear more pronounced in the Austrian case, whereas the British data are notable in that the peripheral North region has similar adoption levels to the metropolitan core; the Austrian periphery would appear to be lagging, particularly in terms of CNC adoption. In Britain it is the industrial heartland of the old iron-based areas and the metropolitan hinterland where technology adoption appears furthest advanced. In both countries the data suggest that for industries such as these the traditional industrial heartland is often a leading area with respect to technology adoption.

The applicability of particular technologies varies between sectors. Alderman et al. (1988) have demonstrated in the British context that, for technologies such as NC and CNC, interindustry diffusion rates vary more than interregional ones. In Table 8 we show that, despite the crude level of sectoral disaggregation employed, differences between the metalworking and machinery sector and the electrotechnical sector are similar in both countries. However, levels of NC and CNC adoption in the Austrian metalworking and machinery sector appear to be relatively lower than in Britain, which may in part reflect the age and size structure of the Austrian sample, but is nevertheless somewhat surprising, given that these are now considered mature technologies (Ray 1984) and it has been argued that CNC in particular is increasingly suited to the operations of the small engineering firm (Dodgson 1985).

Table 8   Adoption rates (percent of establishments) of new technology by industry sector (sources: see Table 3)

Technology	Metroworking and machinery		Electrotechnical and electronic products	
	Austria	Britain	Austria	Britain
NC machines[a]	11.7	33.7	18.2	27.0
CNC machines	16.5	52.4	39.4	59.5
CAD, CAE	9.7	23.3	15.2	43.2
CAM, CAD-CAM	8.7	nd	18.2	nd
Microprocessors	nd	30.7	nd	48.6
Computers for commercial use	73.8	77.3	87.9	89.2

[a] 1981 rate for Britain, notes see Table 7

The most striking sectoral differences, particularly in the British case, occur with respect to CAD adoption. The electrotechnical sector has found CAD to be particularly relevant in relation to printed-circuit-board design, where computerised methods were first developed in the 1960s (Kaplinsky 1984). Note that in the British case these sectoral differences are only statistically significant in the case of CAD and microprocessor adoption.

It was noted above that the corporate structure of Austrian industry is rather different from that of Britain. Table 9 reveals that technology adoption by corporate status also differs. Headquarters (strictly speaking, establishments with control functions) in the British case appear to have a higher propensity to adopt than their Austrian counterparts. In Britain it is the independent (usually small) establishments that are least likely to adopt new technologies, whereas in Austria the branch-plant sector performs comparatively poorly.

Another factor commonly regarded as important in relation to technology adoption is R&D activity. The precise relationship between R&D and technology adoption has yet to be satisfactorily identified. In relation to product innovation its importance is clear (Thwaites et al. 1981), but in relation to process innovations the effect of R&D is frequently confounded with the effect of establishment size, as larger establishments are more likely to support R&D activities. Of the technologies under consideration here, the one that has the closest a priori link with R&D is CAD, as design activities are an intrinsic part of the R&D process.

By concentrating on the *proportion* of employment within the establishment that is engaged in R&D it is at least partially possible to control for the size effect. Careful inspection of Table 10 reveals the inconclusiveness of any evidence for a clear-cut relationship between technology adoption and R&D, particularly in the case of NC or CNC. Care should be taken in interpreting these figures, however, because there are comparatively few establishments with more than 10% of their employees in R&D. Moreover, the largest establishments are unlikely to have the largest proportional levels of R&D, because the absolute numbers involved would be unrealistic.

Table 9　Adoption rates (percent of establishments) of new technology by corporate status (sources: see Table 3)

Technology	Single-plant enterprise		Multiplant establishment			
			Headquarter		Branch	
	Austria	Britain	Austria	Britain	Austria	Britain
NC machines[a]	10.1	22.9	16.7	48.1	15.8	35.0
CNC machines	23.2	40.0	25.0	75.9	10.5	55.3
CAD, CAE	7.2	15.2	16.7	50.0	10.5	24.8
CAM, CAD-CAM	8.7	15.2	18.8	nd	0.0	nd
Microprocessors	nd	19.0	nd	59.3	nd	34.0
Computers for commercial use	72.5	63.8	87.5	94.4	68.4	86.4

[a] 1981 rate for Britain, notes see Table 7

Table 10　Adoption of new technology (percent of establishments) related to the percentage of R&D staff in total employment (sources: see Table 3)

Technology	0%		1-4%		5-9%		≥10%	
	Austria	Britain	Austria	Britain	Austria	Britain	Austria	Britain
NC machines[a]	11.4	19.0	16.4	38.9	16.7	48.1	4.5	18.2
CNC machines	14.3	34.5	20.0	65.1	33.3	51.9	27.3	36.4
CAD, CAE	5.7	15.5	7.3	32.0	20.8	25.9	18.2	45.5
CAM, CAD-CAM	5.7	nd	16.4	nd	4.2	nd	13.6	nd
Microprocessors	nd	15.5	nd	42.1	nd	33.3	nd	54.5
Computers for commercial use	65.7	65.5	90.9	87.3	75.0	74.1	63.6	81.8

[a] 1981 rate for Britain, notes see Table 7

In relation to CAD adoption there is less evidence for a negative trend in adoption as the proportion of R&D staff increases, which is more consistent with expectations. However, there are differences between Britain and Austria in that the major increase in adoption propensity occurs between the 1-4% and 5-9% categories in the Austrian case and between the 0% and 1-4% categories in the British case, indicating that CAD adoption is occurring in establishments with a proportionately lower formal commitment to R&D in Britain than in Austria. Further analysis is required here, because in Table 10 we have not considered informal R&D activities, and a lot depends upon how executives classify R&D staff. CAD systems may be ideal for establishments where there is a lot of routine modification of design and this type of activity may or may not be classed as R&D.

4   Logit Analysis

In the foregoing analysis, we have revealed some consistent patterns of technology adoption between Austria and Great Britain, together with some intriguing contrasts. As noted, however, the differences in the structure of the two samples in terms of size, status, age distribution, and so on limit the extent to which firm conclusions can be drawn. As a first step in overcoming these difficulties, the data were also analysed by means of logit models in order to control for such effects and to identify real differences between the two countries and to establish the extent to which regional variations can be attributed to other factors. Logit modelling is an attempt to overcome the difficulties inherent in bivariate analysis with the rigour of multiple regression modelling for categorical data with a dichotomous response variable (for more details, see Fischer and Nijkamp 1985; Wrigley 1985).

In the simple bivariate analyses reported above no account has been taken of differences in timing and structure of the samples. Before attempting to put the two data sets together it was necessary to remove some of the obvious sources of inconsistency that might otherwise have biased comparative results.

The most serious of these concerns the fact that the British survey was a follow-up survey and that, consequently, no establishments founded after the middle of 1981, the time of the original survey, were included. All British plants are therefore at least six years old (at the time of writing), whereas their Austrian counterparts in some cases are much younger. To overcome this limitation we exclude establishments that started up after 1980. This reduces the size of the Austrian sample to 110 cases.

Definitional differences also mean that some of the technologies referred to above could not be compared analytically. NC was not included in the British 1986 survey on the grounds that it had been largely superseded by CNC and therefore the time periods that are being compared are different. If one bears in mind the aforementioned provisos, three innovations were suitable candidates for analysis: CNC, CAD, and computers for commercial applications. These technologies allow us to examine the three main spheres of manufacturing activity: production, design, and co-ordination, respectively (Kaplinsky 1984). These make up the three dichotomous dependent variables of the form adopted or not adopted. The restricted nature of the British postal survey again limits the number of independent variables available; however, the following were incorporated into the analysis:

(a) location (peripheral region, metropolitan core, metropolitan hinterland, traditional iron-based region);
(b) establishment employment size (natural logarithm);

(c) corporate status (independent, headquarter, branch);
(d) sector (metalworking and machinery, electrotechnical and electronic products);
(e) age (more than fifteen years old, up to fifteen years old);
(f) degree of product diversification [low, high (more than four major product groups)].

In all of the model results that follow, with the exception of the size variable which is continuous, parameter estimates may be interpreted with respect to the reference category. The reference category is a function of the particular parameterisation used by the estimating package (GLIM) and is set to zero. The reference category consists of independent establishments over fifteen years of age in the metalworking and machinery sector that are located in the peripheral region and have a low level of product diversification.

4.1 Single-Nation Models

The first step in the analysis was to compute separate models for each innovation and each country. Table 11 indicates the degree to which establishment characteristics increase or decrease the probability (strictly the logarithmic odds) of adoption of CNC, CAD, and computers in commercial applications. There is no intention that the results presented in this table should in any sense represent 'optimal' models. Rather, the approach is essentially exploratory and the intention is to demonstrate which variables are important and to identify whether the magnitudes and directions of the relationships are similar or otherwise. Although *t*-values are given as well as the parameter estimates, it should be noted that the most reliable way to evaluate the significance of the estimates is through the change in log-likelihood associated with each parameter. For variables with more than two categories the significance of any one parameter will depend on its relationship to categories other than the reference category which is what the *t*-value reflects.

In the case of CNC adoption, it should be clear from Table 11(*a*) that in Britain the dominant factor is the size variable, and locational effects are not significant. The model simplifies to the size effect, the larger the establishment the higher the probability of adoption, and to a possible age effect whereby younger establishments have a lower probability of adoption. In Austria, by way of contrast, there is a strongly negative branch-plant effect and the electrotechnical sector exhibits a higher level of adoption than the metalworking and machinery sector. There are also strong regional effects reflecting the fact that adoption levels are much higher in all areas compared with the periphery (the reference category), but most notably in the metropolitan hinterland and the traditional iron-based region.

Table 11(*b*), on the other hand, indicates that there is very little variability in CAD adoption in Austria. A very low value of $\bar{\rho}^2$ is accompanied by a predictive success of 90%! This is probably because of low overall levels of adoption of

CAD and may have been exacerbated by the removal of younger plants. Only product diversification is a significant factor here; as one might anticipate, greater diversity increases the probability of CAD adoption. In Britain, size is again an important factor and a significant location effect reveals higher levels of CAD adoption in the metropolitan hinterland than elsewhere.

Table 11   Parameter estimates for the single-nation model (*t*-values in parentheses)

Variable	(*a*) CNC adoption		(*b*) CAD adoption		(*c*) Computer adoption	
	Austria	Britain	Austria	Britain	Austria	Britain
Headquarter	-1.26 (-1.88)	0.40 (0.84)	0.90 (1.07)	0.41 (0.83)	-1.59 (-1.70)	0.45 (0.63)
Branch plant	-2.54 (-1.99)	-0.07 (-0.21)	0.09 (0.07)	-0.10 (-0.23)	-2.19 (-2.16)	0.58 (1.40)
Size (logarithm of employment)	0.41 (1.85)	0.82 (5.07)	0.02 (0.06)	0.85 (4.72)	1.47 (3.86)	1.39 (5.30)
Electrotechnical sector	1.46 (2.35)	-0.05 (-0.12)	-0.03 (-0.00)	0.69 (1.50)	0.43 (0.41)	0.82 (1.19)
Metropolitan area: Core	2.00 (1.66)	0.32 (0.75)	-0.26 (-0.26)	0.33 (0.63)	1.68 (1.71)	0.21 (0.37)
Metropolitan area: Hinterland	2.80 (2.28)	0.53 (1.19)	0.24 (0.23)	0.99 (1.94)	2.45 (2.20)	0.20 (0.35)
Traditional iron-based region	3.04 (2.40)	0.35 (0.92)	-0.52 (-0.45)	0.38 (0.83)	-0.29 (-0.34)	0.15 (0.30)
High degree of product diversif.	-1.01 (-1.17)	-0.24 (-0.70)	1.47 (1.76)	0.61 (1.43)	1.17 (0.91)	-0.09 (-0.20)
Age less than fifteen years	-0.19 (-0.30)	-0.69 (-1.49)	0.46 (0.60)	-1.10 (-1.62)	-0.17 (-0.22)	1.14 (1.65)
Constant	-4.85 (-3.26)	-3.45 (-4.61)	-3.03 (-2.19)	-5.98 (-6.00)	-4.41 (-2.91)	-4.68 (-4.47)
$\bar{\rho}^2$	0.14	0.14	0.005	0.19	0.35	0.26

Notes see Table 7

Table 11(*c*) indicates that size is more important with respect to computer adoption than to the other technologies in Austria and here the metropolitan regions have significantly higher levels of adoption than the others. (There is also a significant interaction between size and age of establishment, but this is rather difficult to interpret and may be attributable to a few influential observations.) For the British case, size is again the only significant variable and the removal of all others has negligible impact on the goodness of fit.

It is clear then that, for the most part, regional variations in technology adoption are not significant once other factors have been taken into account, with the notable exception of CNC and computers for commercial use in the Austrian case, where the metropolitan and traditional iron-based areas have a higher probability of adoption than the periphery, and in the case of higher levels of CAD adoption in the metropolitan hinterland in Britain. These observations accord with the suggestion that regional variations are likely to be most pronounced when technologies are in their infancy, but that as diffusion proceeds and approaches saturation level regional convergence is likely to be observed (Alderman and Davies 1990).

4.2 Dual-Nation Models

The single-country model provides a test of within-country variations in technology adoption. The dual-nation model allows us to test formally whether or not there are significant differences between Austria and Great Britain in this respect. This involves the addition of a new independent variable taking the value 1 if the establishment is Austrian and the value 2 if it is British. The reference category in these models consists of the Austrian establishments, with all other characteristics the same as for the single-nation model. The new variable appears first in Table 12, which gives the results of the logit analyses for CNC, CAD, and computer adoption. In these models we are interested primarily in interaction effects that will indicate whether or not there are significant differences between the two countries in terms of the factors associated with the adoption of these technologies.

Table 12(*a*) shows that for CNC adoption there is a strong and significant difference in adoption between the two countries with a much higher probability of an establishment having adopted CNC in Britain than in Austria. The regional effects are similar, although the single-nation model indicated these to be stronger in the Austrian case, and the effect of establishment size is consistent between countries. The major difference is in terms of the corporate-status effect, indicated by a significant interaction term for independent establishments in Great Britain.

The results in Table 11(*a*) provide the clue as to how this should be interpreted. The independent plants in Austria are much more innovative than their corporate counterparts seem to be, whereas in Britain there would appear to be little difference, once the effects of factors such as size have been taken into account.

In Table 12(*b*) the results for CAD adoption show that the difference between the two countries is again significant, but less pronounced. However, the analysis confirms that the effect of establishment size is significant in the British case, but not in the Austrian case as the main effect term for the size variable becomes negligible, whereas the interaction term is significant. In both countries the metropolitan hinterland has the highest levels of adoption, but the effect is not significant, because the nature of these locational contrasts is not consistent: in Britain the peripheral area has the lowest probability of adoption, whereas in Austria it is the traditional iron-based region that has the lowest probability. Product diversification does appear to be positively associated with CAD adoption, possibly because greater diversity demands, ceteris paribus, higher levels of design and draughting activity.

Table 12   Parameter estimates for the dual-nation logit analysis (*t*-values in parentheses) [a]

Variable	(a) CNC adoption		(b) CAD adoption		(c) Computer adoption	
	ME	MI	ME	MI	ME	MI
Great Britain	1.98 (5.24)	3.06 (5.73)	1.27 (2.70)	-2.94 (-2.17)	-0.23 (-0.50)	2.06 (2.63)
Headquarter	-0.10 (-0.27)	-1.36 (-2.52)	0.53 (1.27)	0.50 (1.19)	-0.18 (-0.36)	-1.47 (-2.03)
Branch plant	-2.29 (-0.98)	-1.96 (-2.20)	-0.04 (-0.10)	-0.13 (-0.33)	0.24 (0.66)	-1.43 (1.92)
Size (logarithm of employment)	0.64 (5.36)	0.64 (5.34)	0.58 (4.22)	0.03 (0.14)	1.24 (6.17)	1.64 (6.71)
Electrotechnical sector	0.30 (0.88)	0.42 (1.20)	0.46 (1.24)	0.51 (1.32)	0.44 (0.54)	0.60 (1.09)
Metropolitan area: Core	0.41 (1.12)	0.43 (1.19)	0.12 (0.27)	0.26 (0.85)	0.74 (1.60)	2.35 (3.11)
Metropolitan area: Hinterland	0.82 (2.18)	0.83 (2.14)	0.85 (1.92)	0.87 (1.89)	0.56 (1.17)	2.45 (3.10)
Traditional iron-based region	0.69 (2.08)	0.68 (2.04)	0.25 (0.62)	0.33 (0.79)	-0.07 (-0.17)	-0.06 (-0.15)
High degree of product diversification	-0.46 (-1.52)	-0.46 (-1.51)	0.62 (1.70)	0.84 (2.15)	-0.02 (-0.04)	0.08 (0.18)
Age less than 15 years	-0.24 (-0.68)	-0.32 (-0.88)	-0.37 (-0.79)	-0.45 (-0.95)	-0.09 (-0.22)	3.73 (2.49)
Independent plant in Great Britain		-1.99 (-3.21)		0.80 (3.12)		-1.94 (-2.49)
Metropolitan area plant in Great Britain						-2.35 (-2.98)
Size by plant less than 15 years old						-0.93 (-2.34)
Constant	-4.59 (-6.94)	-3.95 (-5.90)	-5.95 (-6.87)	-3.13 (-2.87)	-3.72 (-5.09)	-5.60 (-5.30)
$\bar{\rho}^2$	0.16	0.18	0.16	0.18	0.24	0.30

[a] *ME* main effects model; *MI* model with interactions, notes see Table 7

Table 12(*c*) reveals that the model for computer adoption is by far the most complex, with three significant interaction terms. Overall levels of adoption are similar between the two countries, but independent plants in Great Britain are less likely to have adopted than their counterparts in Austria. Regionally, establishments located in either the metropolitan hinterland or the core in Austria are proportionately more likely to have adopted computers for commercial applications than in the equivalent areas in Britain. Young establishments also appear to be more innovative, but the interaction term between size and age indicates that this is less true the larger the plant. This interaction was only observed in the Austrian data and does not appear particularly meaningful. It is possible that the age effect is too crudely defined and should ideally be treated as a continuous variable. Moreover, by definition, all establishments are at least six years old in these models.

By and large, sectoral differences in this dual-nation model are not significant, although the electrotechnical sector is more innovative in terms of CNC adoption in Austria than the metalworking and machinery sector. The sectoral breakdown used here is very crude, however, and the problems associated with matching sectoral classifications were referred to earlier.

5   Constraints on Adoption

The evidence presented here would seem to provide some fairly conclusive evidence that the adoption of new technology in Austrian firms is some way behind that in Britain, notwithstanding the differences in the characteristics of manufacturing industry between the two countries. Identification of the constraints on adoption is obviously an important objective, both from the perspective of individual companies and from a policy point of view.

Some further evidence from the two surveys sheds some light on the major constraints to adoption as expressed by industry executives. In the Austrian case the problems of lack of finance and a lack of suitably qualified staff topped the list (about 36% of establishments). In over 95% of cases, manufacturers called upon internal funds. Bank finance and government assistance was used by a third of respondents only. Comparable figures for the British case are not easily extracted, but corporate establishments relied very heavily on internal or company group funds to support technology adoption, whereas for independent establishments bank finance was more important (Thwaites et al. 1982).

In Britain the dominant constraint appears to be less the lack of finance per se, than the inability to justify the investment. A major constraint on investment in new technology for branch plants in particular is the requirement to demonstrate a very rapid payback (Alderman and Thwaites 1987) and this becomes increasingly difficult the more sophisticated the technology. This is one obvious reason for the slow rate of uptake of new forms of manufacturing technology, such as flexible manufacturing systems (FMS), which remain the preserve of the larger establishments and enterprises (see Bessant and Hayward 1986).

In 1981 the British survey found the lack of suitably qualified staff to be a comparatively minor problem, although this may well be changing, particularly with serious shortfalls foreseen in the information-technology field. Other evidence suggests that it is less likely to be the adoption of technology that is constrained than the successful implementation and operation of the technology once it has been adopted (Alderman and Thwaites 1987).

In the Austrian case the most important sources of information concerning innovation activities were trade journals, sales literature, and exhibitions. Although these were also revealed to be important sources in the British survey, manufacturers' demonstrations and visits by suppliers were considerably more so. It is possible that this is a reflection of the different sizes of domestic market.

Britain is likely to have more equipment manufacturers and suppliers than Austria and a greater reliance by Austria on imports may account for a higher use of exhibitions as important information sources. It is interesting in this context that ÖIAG is currently undergoing a major restructuring, in which it is aimed to secure jobs partly through increasing R&D efforts and gaining access to foreign technologies and products, and this will entail closer links between Austria and the EC.

6 Conclusions

In this paper, we have reported results of an attempt to compare regional and national innovation activity in the Austrian and British contexts, with the use of survey data obtained from a broadly similar group of manufacturing industries. We have demonstrated that significant differences in the structure of industry in the two countries make comparison an extremely difficult exercise. Some initial attempts at controlling for differences in establishment characteristics between the two countries were made through the use of logit analysis.

The results achieved thus far seem to suggest that the Austrian metalworking and machinery sector and the electrotechnical sector, are lagging behind their counterparts in Great Britain in the adoption of manufacturing process technologies, although the use of computers in the commercial sphere is as advanced as in Britain, if not more so. To a large degree these findings arise as a consequence of a younger age and smaller size structure of establishments within the Austrian sample and it is not surprising to discover that the major constraints on adoption in the Austrian sample were lack of finance and lack of suitably qualified staff, which are both typical problems for young, small establishments.

The results of the logit analysis reveal that variations between the four regional types in Britain, to the extent that they exist at all, are largely attributable to different structural characteristics, such as size, ownership, sectoral composition, and so on. Indeed, in the British case establishment size is the dominant factor associated with technology adoption and the only consistently significant one. In Austria regional differences still remain after controlling for these factors, suggesting more deep-seated problems with respect to technology adoption for the peripheral areas. Adoption in Britain has probably proceeded sufficiently far that we are now observing regional convergence in adoption levels.

An intriguing question arises from the finding that the independent establishments in Austria appear to be relatively more innovative than those which are part of larger enterprises, which is in contrasts to the experience in Britain. To the extent that Austria experiences problems of a lack of innovativeness in terms of new manufacturing-process technology it appears to have more to do with larger enterprises and the poor performance of branch plants than it does with the difficulties usually experienced by small independent firms.

Acknowledgements: An earlier version of this paper was presented at the 29th European Congress of the Regional Science Association, Cambridge, 29 August to 1 September 1989. The authors would like to acknowledge the assistance of Gottfried Menschik, who carried out the original computer analysis of the Austrian data, and Kieran McInerney and Trisha Gillespie for help with the British Survey in 1986/87. All views expressed are those of the authors alone and should not be taken to represent those of any of the funding bodies mentioned.

References

Alderman N. and Davies S. (1990): Modelling regional patterns of innovation diffusion in the UK metalworking industries, *Regional Studies* 24 (6), 513-528

Alderman N. and Thwaites A.T. (1987): New technology and its implications for training in the engineering industry, Final Report to the Engineering Industry Training Board, CURDS, University of Newcastle upon Tyne, Newcastle upon Tyne

Alderman N., Davies S. and Thwaites A.T. (1988): Patterns of innovation diffusion, Technical Report, CURDS, University of Newcastle upon Tyne, Newcastle upon Tyne

Bessant J. and Hayward B. (1986): Flexibility in manufacturing systems, *Omega* 14, 465-473

Brugger E.A. and Stuckey B. (1987): Regional economic structure and innovative behaviour in Switzerland, *Regional Studies* 21, 241-254

Commission of the European Communities (1991): *The Regions in the 1990s, Fourth Periodic Report on the Social and Economic Situation and Development of the Regions of the Community*, Office for Official Publications of the European Communities, Luxembourg

Coombes M.G., Dixon J.S., Openshaw S. and Taylor P.J. (1982): Functional regions for the population census of Britain. In: Herbert D.T. and Johnston R.J. (eds.) *Geography and the Urban Environment: Progress in Research and Applications. Volume* 5, Wiley, Chichester, Sussex, pp. 63-112

Davies S. (1979): *The Diffusion of Process Innovations*, Cambridge University Press, Cambridge [MA]

Dodgson M. (1985): New technology, employment and small engineering firms, *International Small Business Journal* 3, 8-29

Fischer M.M. (1990): The micro-electronics revolution and its impact on labour and employment. In: Cappellin R. and Nijkamp P. (eds.) *The Spatial Context of Technological Development*, Avebury, Aldershot, Hants, pp 43-74

Fischer M.M. and Menschik G. (1994): *Innovationsaktivitäten in der österreichischen Industrie. Eine empirische Untersuchung des betrieblichen Innovationsverhaltens in ausgewählten Branchen*, Abhandlungen zur Geographie und Regionalforschung 3, University of Vienna, Vienna

Fischer M.M. and Menschik G. (1991): Innovation und technologischen Wandel in Österreich, *Mitteilungen der Österreichischen Geographischen Gesellschaft* 133, 43-68

Fischer M.M. and Nijkamp P. (1985): Developments in explanatory discrete spatial data and choice analysis, *Progress in Human Geography* 9, 515-551

Freeman C. (1974): *The Economics of Industrial Innovation*, Penguin Books, Harmondsworth [UK]

Gibbs D.C. and Thwaites A.T. (1985): The international diffusion of new technology in manufacturing industry: A comparative study of Great Britain, the USA and West Germany, Paper presented to the Institute of British Geographers, CAG, Symposium on "Technical Change in Industry-Spatial Policy and Research Implications", Swansea; copy available from CURDS, University of Newcastle upon Tyne, Newcastle upon Tyne

Jacobsson S. (1985): Technical change and industrial policy, the case of computer numerically controlled lathes in Argentina, Korea and Taiwan, *World Development* 13, 353-370

Kaplinsky R. (1984): *Automation: The Technology and Society*, Longman, Harlow, Essex

Kleine J. (1982): Location, firm size and innovativeness. In: Maillat D. (ed.) *Technology: A Key Factor for Regional Development*, Georgi, St. Saphorin, pp. 147-174

Nabseth L. and Ray G.F. (1974): *The Diffusion of New Industrial Processes*, Cambridge University Press, Cambridge [MA]

Northcott J. (1985): *Microelectronics in Industry: An International Comparison*, Policy Studies Institute, London

Northcott J. and Rogers P. (1984): *Microelectronics in British Industry: The Pattern of Change*, Policy Studies Institute, London

Quévit M. (ed.) (1991): *Regional Development Trajectories and the Attainment of the European Internal Market*, Groupe de Recherche Européenne sur les Milieux Innovateurs, 90 rue de Tolbia, 75634 Paris, Cedex 13

Ray G. (1984): The diffusion of mature technologies, OP-36, National Institute for Economic and Social Research, Cambridge University Press, Cambridge [MA]

Rees J., Briggs R. and Oakey R.P. (1984): The adoption of new technology in the American machinery industry, *Regional Studies* 18 (6), 489-504

Thwaites A.T. (1978): Technological change, mobile plants and regional development, *Regional Studies* 12 (4), 445-461

Thwaites A.T., Edwards A. and Gibbs D.C. (1982): Interregional diffusion of production innovations in Great Britain, Final Report to the Department of Trade and Industry and the Commission of the European Communities; CURDS, University of Newcastle upon Tyne, Newcastle upon Tyne

Thwaites A.T., Oakey R.P. and Nash P.A. (1981): Industrial innovation and regional development, Final Report to the Department of the Environment; CURDS, University of Newcastle upon Tyne, Newcastle upon Tyne

Tödtling F. (1990): Regional differences and determinants of entrepreneurial innovation. Empirical results from an Austrian case study. In: Ciciotti E., Alderman N. and Thwaites A.T. (eds.) *Technological Change in a Spatial Context*, Springer, Berlin, Heidelberg, New York, pp. 260-284

Wrigley N. (1985): *Categorical Data Analysis for Geographers and Environmental Scientists*, Longman, Harlow, Essex

3 Technology, Organisation and Export-driven Research and Development in Austria's Electronics Industry

with *L. Suarez-Villa*

Over the past two decades Austria's export-driven electronics industry has experienced a progressive territorial distribution that has substantially decentralised production and employment. Nevertheless, the capital region's concentration has provided many advantages to R&D-intensive establishments through subcontractual opportunities and better access to advanced research and production skills. This paper analyses the relationship between R&D, territorial location and the most important organisational characteristics in Austria's electronics industries. An assessment of operational motivations, based on establishment-level survey data, is followed by a factor analysis that reveals the main organisational dimensions. Statistical analyses of the association between R&D intensity, territorial location and the organisational factors are then expanded to consider subcontracting and just-in-time production methods. Two-way subcontracting, whereby firms both subcontract part of their production out and are in turn contracted by others, is found to be prevalent. Such arrangements are thought to help firms specialise and avoid implementing costly production techniques whilst helping save capital and resources that can be reinvested in R&D. These analyses provide important insights on the association between R&D and subcontracting, just-in-time production and on the advantages of skilled production labour and plant size for research-intensive manufacturing establishments.

Introduction

Austria's small land-locked territory and internal market, and its peripheral location bounding Eastern Europe's closed borders, became major disadvantages over much of the post-war era. These obstacles were compounded by the decline of traditional industries and limited access to capital markets that increased its isolation and fiscal burdens. Despite these difficulties, Austria's promixity to the larger Western European consumer markets, its favourable wage cost differentials and its high labour skills allowed it to gain a significant position in electronics manufacturing over the past two decades.

The rise of Austria's electronics industry had much to do with its location outside the European Common Market, as German and West European industrial restructuring deepened, and lower-cost production sites with high skills became attractive alternatives. Significant export market niches were developed in such

industrial and component electronics sectors as telecommunications, audio technology and microprocessors, whilst investment in new technology and R&D
increased the innovative capacity of both domestic and foreign-owned electronics
firms. These factors raised Austria's profile as an electronics producer, as Western
Europe sought to limit East Asian penetration in electronics production and
markets (see, for example, Dosi et al. 1988; Jacquemin and Sapir 1989; Amsden
1990; Best 1990; Fischer and Menschik 1991, 1994; Mattsson and Stymne 1991;
Amin et al. 1992; Jones and Schröter, 1992).

Austria's competitive insertion in Western European electronics production
required a significant enhancement of innovative capabilities to develop export
market niches through specialisation and to keep abreast of rapid technological
change. Few other industries have advanced technologically as rapidly as
electronics, with major breakthroughs that have often followed each other by a
matter of months (see, for example, Braun and MacDonald 1982; Ernst 1983;
Soete and Dosi 1983; Hughes 1986; Pavitt 1986; Swann 1986; Dorfman 1987;
Howell et al. 1987; Miles 1988; Todd 1989; Suarez-Villa and Han 1990a; Fischer
1990; Henderson 1991; Mathias and Davis 1991). This rapid technological pace
has required the development of strong R&D capabilities to establish markets for
the more technologically advanced goods, especially in the industrial and
component electronics sectors, and to remain internationally competitive over
longer periods of time.

Differences in the rate of technological change, technological sophistication
and, consequently, in R&D intensity can vary widely between the various
electronics sectors and even among various types of products. This requires the
careful targeting of R&D activities to develop innovations for specific goods that
can establish their own market niches within a reasonably short time. In Austria's
case, R&D has been most successful in the development of industrial electronics
goods, such as audio and broadcasting technology, and telecommunications equipment, and in the development of components and accessories for these product
segments. Closely linking R&D activities to specific product types and their
marketing objectives has also helped to develop endogenous R&D capabilities
among export-oriented firms in the industrial electronics sector.

Whilst an overarching emphasis on exports has promoted the spatial dispersion
of branch plants in some of the more important producer nations, it remains
unclear to what extent R&D-intensive electronics production has decentralised
(see, for example, Suarez-Villa and Han 1990b, 1991). Substantial differences in
R&D emphasis could occur between hinterland and metropolitan establishments if
significant externalities for R&D and product innovation favour concentration.
More limited access to the kinds of skills and external functions that support R&D
may be a serious disadvantage for hinterland locations with smaller industrial
bases and fewer opportunities for localised transactions. Favourable production
and siting cost differentials in the hinterland may therefore not be enough to offset
the advantages of concentration for R&D-intensive establishments for which
timely innovation, specialisation and external transactions are important prerequisites.

Locating or remaining in places with significant concentrations of similar firms may be essential to develop strong R&D capabilities if externalisation results in resource savings that can be reinvested in R&D. This is particularly important where limited access to venture capital has severely constrained technological development and adoption (Fischer and Menschik 1991; Alderman and Fischer 1992). Finding more and better opportunities to externalise functions through concentration may therefore be a significant pecuniary advantage for innovation and R&D investment, where cost reduction becomes an important means of generating investment capital.

Externalising functions through subcontracting may also help save skilled labour for in-house activities that are closely linked with R&D, particularly whenever personnel can be interchangeably employed in production and R&D. This is important where training is costly and where much of it has to be borne internally by firms as part of in-house apprenticeship or upgrading processes. Such skills and training are usually essential for the adoption of new technologies, and their scarcity or high costs can be effective barriers to production and to qualitative product development (Fischer and Menschik 1994; Alderman and Fischer 1992). The concentration of establishments may therefore also help improve access to skilled labour, especially where an already established manufacturing culture and training centres can provide significant economies.

Complexities in the externalisation of functions, such as *two-way* sub-contracting, may be inherent to the spatial concentration of R&D-intensive firms. By taking advantage of diverse local transactional and co-operative possibilities, firms engaging in two way subcontracting might externalise part of their pro-duction whilst they are, at the same time, subcontracted by other firms for some production tasks. The motivation for such arrangements might be both internal and external, as firms strive to utilise idle productive capacity during periods of lower demand without sacrificing ongoing or agreed-upon transactions with their suppliers. Such arrangements might simultaneously assist other firms in meeting their production objectives during times of stress or of high market demand, as a way of eliciting reciprocity or as a trust-building strategy. If R&D is a major beneficiary of such arrangements, through the resource savings and reinvestment that may occur, it would underline the paradoxical coexistence of co-operation and trust with competition, as an important aspect of the process of industrial concentration and vertical disintegration. R&D is by far the most competitively oriented and closely held aspect of the industrial firm, given its crucial role in determining the timeliness of market entry and of long-term performance.

Firms that concentrate may therefore co-operate in ways that reduce costs whilst at the same time competing on quality and technique, although perhaps less so on price or for the same market segment (see, for example, Piore and Sabel 1984; Becattini 1989; Sabel 1989; Brusco 1990; Porter 1990; Hansen 1992). R&D activities, competitively oriented as they are, may nevertheless be the most important beneficiaries of such co-operative arrangements, given their capital-intensive character and the contribution that co-operation can make towards the internal generation of capital. Firms may also seek co-operative arrangements as part of a broader risk-reduction strategy, where R&D resources are generated internally, whilst venture capital from external sources is sought to finance the more visible and less risky marketing, productive process or overhead capital investments. The literature on R&D and corporate strategy has traditionally acknowledged research to be the riskiest and most uncertain aspect of organisations, where its scope and objectives tend to be carefully followed by the upper echelons of management (see, for example, Marschak et al. 1967; Mansfield 1971; Freeman 1974; Donaldson and Lorsch 1986; Hounshell and Smith 1988; Håkansson 1989; Thwaites and Alderman 1989; Malecki 1991; Pornschlegel 1992). The interaction of location, corporate strategy and R&D resource allocation provides fertile grounds to analyse the extent to which external transactions may be significant determinants of R&D and corporate innovation.

Firms engaging in vertical disintegration also stand to benefit greatly from concentration by gaining access to a more finely detailed local division of labour, where specialisation may help reduce costs whilst achieving qualitative improvements in production (see, for example, Teece 1980; Piore and Sabel 1984; Williamson 1985; Scott 1988; Chandler 1990). Clearly, the more finely detailed division of labour found in large concentrations of similar firms provides a wider range of localised choices and possibilities to specialise by externalising functions through subcontracting. The wider range of possibilities may also make it easier for firms to avoid the costly adoption of new production techniques, whenever some functions can be shifted to existing or newly found suppliers. Occurring through simple or pecuniary externalities or through industrial district effects, the benefits derived from such arrangements may greatly determine the research intensity of firms.

This study will explore the relationship between R&D, territorial location and the organisational characteristics that have shaped Austria's electronics industry. The territorial distribution of electronics establishments and employment will be considered first, to show how an emphasis on exports transformed the regional division of labour for electronics production. The primary motivations for establishment operation in both the hinterland and the capital region will be discussed in a subsequent section, to determine whether any territorial differences in operational objectives can be found. This will be followed by an analysis of the main organisational characteristics of Austria's electronics industry, based on a representative sample of establishments. The relationship between the main organisational factors and R&D will then be explored statistically, to determine the extent to which spatial concentration affects R&D outlays in electronics establishments. Finally, the consideration of subcontracting and of just-in-time production methods will provide insights on how these arrangements affect R&D intensity, and their relationship with organisational scale, skills and production processes, as firms seek to maintain their research capabilities.

2 Exports and Territorial Distribution

Exports account for nearly 85% of Austria's electronics production, making this industry one of its most important sources of foreign earnings (Fachverband der Elektro- und Elektronikindustrie 1992). Over 75% of Austria's electronics exports are sold in Western Europe, where they have gained significant market niches and have successfully competed with East Asian imports. Austria's export performance was helped by the depth of Western European industrial restructuring since the late 1960s. In the late 1970s and throughout the 1980s, Austria's electronics industry was bolstered by Southern Germany's emergence in high-technology manufacturing, and by West Germany's continuous industrial re-structuring. German capital looked for lower-cost production sites in the Common Market's periphery as a way to offset East Asian competition, whilst Austria's domestic manufacturers found new marketing opportunities in the West (see, for example, Molle and Klaassen 1985; Fischer and Nijkamp 1987, 1988; Jacquemin 1987; Braun and Polt 1988; Ratti 1988; Lane 1989; Cappellin and Nijkamp 1990; Matzner and Wagner 1990; Bade and Kunzmann 1991; Cheshire 1991; Altmann et al. 1992; Fritsch 1992; Quévit 1992).

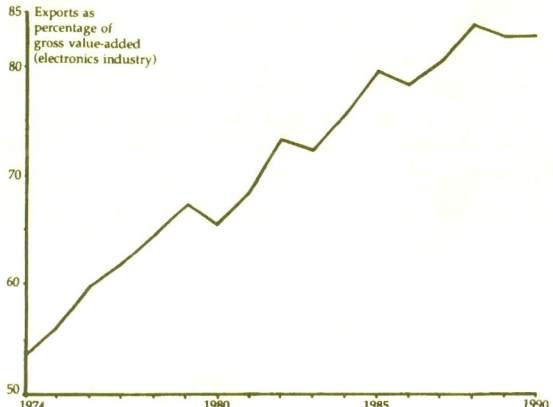

Figure 1   The growing importance of exports in Austria's electronics industry, 1974–90. Source: Based on data obtained
from Fachverband der Elektro- und Elektronikindustrie,
1974-91

The steady and rapid growth of exports in Austria's electronics industry, shown in Figure 1, is remarkable in that it occurred during periods of global recession. Thus, despite the economic downturns of the middle and late 1970s, and the early 1980s, Austria's electronics exports increased by approximately 20% as a proportion of gross output between 1974 and 1982 (Österreichisches Statistisches Zentralamt, 1969–89). By the late 1980s, exports accounted for 82% of the electronics industry's gross output, reflecting an aggressive expansion that was successful even by the standards of the East Asian producers (see, for example, Amsden 1990; Suarez-Villa and Han 1990a). Lacking a favourable or sizeable home market, Austria's experience defies one of the central assumptions of contemporary trade theory regarding export performance (see, for example, Krugman 1991). A strategic emphasis on niche market development may account for much of this performance, and may need to be taken into account in future analysis of small nation manufacturing trade and exports.

Austria's significantly lower manufacturing wage costs, averaging 35% below West Germany's throughout the 1970s and 1980s, stimulated much branch plant investment as well as domestic-owned but foreign-contracted assembly operations (Maier and Weiss 1986; Österreichisches Institut für Wirtschaftsforschung 1991; Fachverband der Elektro- und Elektronikindustrie 1992). Relatively high skills, adequate infrastructure, lower land costs and a shared language were also significant factors attracting German capital, particularly in the hinterland regions. Austria's geographical proximity to West Germany, Europe's largest electronics producer, was advantageous, providing easy access for transborder subsidiary transactions and subcontracting. In Western Austria, for example, it became quite feasible to manage branch operations directly from a firm's German base, since some locations involved only a relatively short commute from the German border and from major industrial centres, such as Munich. Bavaria's emergence as Germany's most important region for high-technology thereby became a major source of capital and technology for branch plant investment.

Government participation in Austria's electronics industry is insignificant today, despite a history of heavy state ownership in various industrial sectors, such as steel and other heavy industries. Social democratic governments in the 1970s, despite their attempts to maintain industrial employment, practically eliminated state participation in electronics, divesting their holdings in this industry. An important benchmark in this effort was the privatisation of Siemens Austria, one of the most important industrial concerns, in 1971. This was radically different from the situation after the end of the Second World War, when almost half of all electrical and electronic manufacturing firms were government owned (Surholt 1984).

Austria's protectionist regulation with respect to Western European electronics goods has been practically non-existent, at least since the emergence of the electronics industry as a significant exporter (Austria Offert 1991). As a founding member of the European Free Trade Association (EFTA), neither quotas nor customs duties were imposed on Austria's imports from other EFTA member nations since 1967. Similarly, duties and quotas on goods imported from the European Community [EC] were removed after the bilateral EFTA–EC agreement negotiated in 1972, and fully implemented in 1977. Duties on imports from non-EFTA and non-EC nations have remained at relatively high levels, however, with tariff rates of as much as 20%. Exemptions are nevertheless made for imports from non-EFTA and non-EC nations that incorporate Austrian-made components, charging duties as low as 4%. Because of Austria's trade with Western Europe, a liberal trade orientation was seen as an important prerequisite for increasing exports to continental markets (Braun and Polt 1988; Bayer 1992; Matis 1994).

Austria has had virtually no specific export support policies and programmes for the electronics industry, although some general mechanisms have benefited electronics, as they have most other manufactured exports. For example, many export loans and credits have been financed by the Central Bank (Österreichische Kontrollbank), especially when other sources of credit could not be found. The rates charged were usually at levels similar to those of the major financial markets, however. A second general instrument has been the network of trade offices managed by the Federal Chamber of Commerce (Aussenhandelsstellen der Bundeswirtschaftskammer), assisting small and medium sized firms in identifying export opportunities. Neither of these mechanisms were targeted to serve the electronics industry in particular, and their efforts have more often than not aided firms in other industrial sectors (Austria Offert 1991).

Similarly, federal and local support for technical and labour skills training has been rather general in scope, and has tended to favour other industries perhaps even more than electronics. This has made it necessary for many electronics firms to implement their own in-house apprenticeship and technical training activities, to address their specific needs and shortages. Large firms typically have more resources for such efforts, whilst small and medium sized enterprises tend to resort to subcontracting and to other contractual or co-operative arrangements, as a way to avoid the costs of in-house technical training. With the shifting emphasis from consumer goods production towards industrial and components production by the middle 1970s, the need for independent or in-house training increased, as small-batch schedules became more common with skill requirements that often varied between production runs (Austria Offert 1991).

The changing sectoral emphasis of Austria's electronics industry required the development of stronger research capabilities over the 1970s and 1980s. R&D-intensive industrial and component electronics goods had the best possibilities for developing export market niches, as Western Europe sought to stave off any substantial market penetration by East Asian imports. The new emphasis produced important technological benchmarks for Austria's electronics industry, such as the introduction of the first European-made 256K microchips in 1985. By the late 1980s, a major, domestic-owned, audio electronics manufacturer (AKG) had established claims to over 1,400 patents of which 300 were fundamental inventions in audio technology. The amount of investments in R&D was equally impressive, with some domestic-owned firms, such as Schrack AG (specialised in tele-communications equipment), typically investing over 14% of gross revenues in research. Siemens AG Österreich, Austria's third largest manufacturing concern, continuously employed over 20% of its personnel in R&D activities by the late 1980s. Such advances exacted a high cost from most firms, as resources had to be generated through the internal rationalisation of tasks and priorities and, whenever possible, by externalising functions through subcontracting (Bundeskammer der Gewerblichen Wirtschaft 1990; Austria Offert 1991; Polt 1992; Fischer and Menschik 1994; Fischer and Schuch 1994).

Export market targeting strategies have required R&D efforts aimed at incremental or continuous innovation, where practical learning occurs jointly between R&D activities and production. This has reinforced the need for highly skilled labour, with often substantial in-house training and learning, and with occasional dual service in production and research. The innovations that result thus tend to have a more firm specific character and are less basic, or less applicable in other sectors, given their more limited range of potential usage. This characteristic makes co-operative (rather than competitive) subcontracting safer, from a proprietary standpoint, since innovations become easier to safeguard whenever they are disclosed to subcontractors. Incremental innovation therefore motivated a need for concentration, where skilled labour, subcontracting and other transactional arrangements are generally more accessible.

From one perspective, therefore, it is obvious that the emphasis an R&D and export market targeting promoted the concentration of technologically advanced production. However, a look at the evolving distribution of Austria's electronics establishments reveals a strong and persistent tendency towards territorial dispersion (see Figure 2). The dualistic character of this phenomenon has been underpinned by the growth of branch plants, or of domestic-owned but foreign-contracted establishments in the hinterland. Branch operations tended to establish direct export and import linkages with their parent plants or foreign contractors, with low local and regional embeddedness. In contrast, the concentration of electronics establishments in the capital region (Vienna and surrounding areas in Lower Austria and Burgenland) provided opportunities for R&D and sub-contracting that were not available in the hinterland's locales (see, for example, Stöhr 1986; Tödtling 1987, 1990; Sheppard et al. 1990; Fischer and Menschik 1994). As a result, the direct import and export transactive intensities of hinterland and capital region electronics establishments, shown in Table 1, are remarkably different.

Table 1   Export and import intensity of Austria's electronics establishments

	Capital region (% of plants)[a]	All other regions (% of plants)[a]
Direct exports per plant		
over 60% of production	19	62
over 30% of production	52	83
Direct imports per plant		
over 60% of inputs	10	50
over 30% of inputs	48	72

[a] refers to the total number of establishments in each of the two regional categories: (1) capital region (Vienna, Burgenland and Lower Austria; $N = 23$); and (2) all other regions ($N = 35$). Estimates do not include indirect exports and imports occurring through linked producers or services. All estimates are based an data obtained through the authors' survey, 1992.

The failure to externalise functions locally, which accounted for the much higher direct import and export transactions of the hinterland's plants, is a reflection of larger processes shaping production and investment. Most of the hinterland's establishments were, from the start, part of a spatial division of labour that was becoming increasingly international, and that was determined by corporate restructuring strategies (see, for example, Watts 1981; McCalman 1989; Best 1990; Sheppard et al. 1990; Jessop et al. 1991; Mason et al. 1991; Veltz 1991; Amin and Malmberg 1992; Amin et al. 1992; Buckley and Ghauri 1993; Cappellin and Batey 1993; Phelps 1993). Intra-corporate transactions and trade motivated much inward investment, by seeking lower-cost sites in Austria, and constrained processes of vertical disintegration that could lead to a localised externalisation of functions. Thus, branch plant investment became part of a refined corporate-wide division of labour, where each establishment assumes a particular 'fit' in the global operational scheme of the parent organisation. Such transborder, intra-corporate trade has indeed become a vital element in the growth and financial well-being of multilocational firms. For some firms, intra-corporate networks of branch plants have led to decentralised, or more spatially dispersed, patterns of production. Recent research on branch plants in the United Kingdom has revealed the extensive character of such networks, and the spatially dispersed profile of their operations (see, for example, Hargrave 1985; Dunning 1988; Mason et al. 1991; Phelps 1993; Turok 1993a, 1993b).

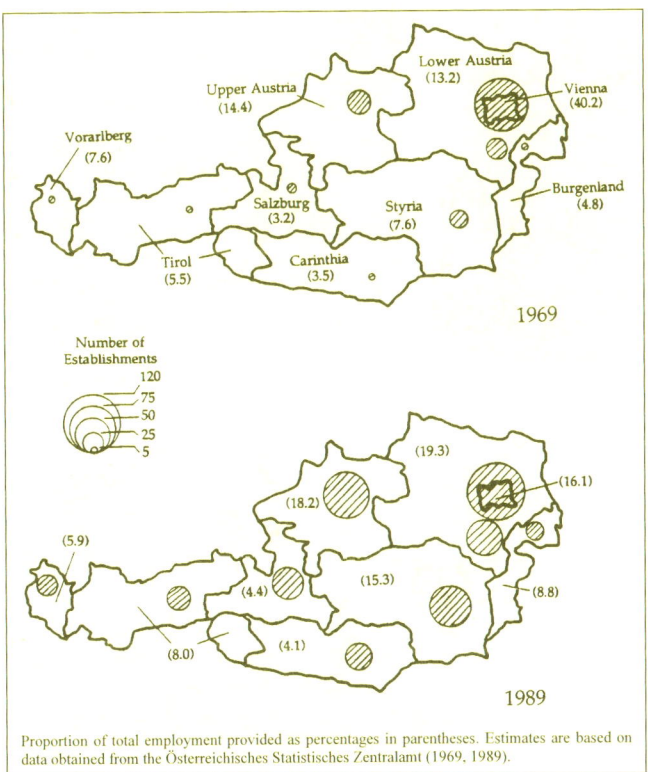

Figure 2   Distribution of Austria's electronics establishments: 1969 and 1989

The lack of local embeddedness found in UK branch plants applies similarly to Austria's hinterland establishments (see, for example, Turok 1993a). In Austria's case, however, the lack of embeddedness is more acute, given the smaller clusters of establishments and the proximity of most hinterland locales to the parent companies or contractors across the German border. The local milieu of the hinterland's cities also limited the possibilities for embedding production; since most places are small or medium sized, they tend to be oriented towards services, and in most cases they have limited experience with industrial enterprise and production (see, for example, Maier and Tödtling 1986; Sheppard et al. 1990). The decline of old-line heavy industries in Styria, for example, also limited the possibilities for embedding inward investment, since branch operations tended to favour 'new spaces' with none of the encumbrances of a bygone industrial era. The opening of borders with Eastern Europe seems to have had little effect on this situation, at least in the case of Styria (Steiner and Sturn 1992).

Austria's difficult topography restricts north-south communications, and only a relatively small proportion of its land surface is potentially suitable for industrial development (Sheppard et al. 1990). Such obstacles are overcome by a good transport and communications infrastructure but others, such as the availability of capital for local firm start-ups, labour training and the development of local managerial skills and education, have been more intractable. It should therefore not be surprising that the disembedded character of most of the hinterland's branch operations has produced few local developmental effects. Despite the proliferation of establishments, the growth of employment in electronics manufacturing in the hinterland has been modest. In Upper Austria and Carinthia employment growth in electronics remained virtually stagnant, in absolute terms, between 1969 and 1989, whilst it actually declined by 41% in Vorarlberg over the same period, despite an increase in the number of production establishments from six to 20 (Österreichisches Statistisches Zentralamt 1969, 1989). Of all the hinterland regions, Styria and Salzburg fared best in absolute employment growth with increases of over 45% and 28% between 1969 and 1989, respectively (*ibid.*). These increases are hardly the sort that can be expected to provide significant local developmental effects, despite the rise in the number of establishments and production capacity.

Despite its low local developmental effects, the growth in establishments and productive capacity in the hinterland transformed the spatial distribution of Austria's electronics industry (see Table 2). In less than two decades, the capital region's relative share of employment and production facilities was significantly diminished. In 1969, Vienna and its surrounding regions (Burgenland, Lower Austria) concentrated 63% of all electronics plants and over 58% of employment. Twenty years later, the spatial distribution of the electronics industry was virtually reversed. The changes were more dramatic when only the capital is considered; Vienna's proportion of employment declined from 40% to 16% of the national totals between 1969 and 1989, whilst its share of establishments decreased from 50% to 28% over that period. This benefited Burgenland and Lower Austria through the growth of ex-urban establishments and employment as firms dispersed from Vienna towards adjacent areas and communities (Maier and Tödtling 1986; Österreichisches Statistisches Zentralamt 1989; Kubin and Steiner 1992).

Table 2   Changes in the spatial distribution of Austria's electronics industry, 1969–1989

	% of employment		% of establishments	
	(a)	(b)	(a)	(b)
1969	58.2	41.8	63.0	37.0
1989	44.2	55.8	41.2	58.8

(a) capital region (Vienna Burgenland and Lower Austria); (b) all other regions; all estimates are based an data obtained from the Österreichisches Statistisches Zentralamt (1969, 1989).

This inversion process is important, given Austria's small territory and its difficult topography throughout most of its hinterland. In small nations, the propensity to maintain industrial concentration is usually greater than in countries with sizeable territories. Primate metropolises in small nations typically concentrate a vast majority of industrial employment and establishments in most sectors. In Austria's case, however, branch plant investments and most of the hinterland's proximity to Germany and Western Europe promoted a dispersion trend that was similar to those occurring in larger Western European nations (see, for example, Oakey et al. 1980; Keeble et al. 1983; Aydalot 1984; Schackmann-Fallis 1989; Milne 1990; Lawton Smith 1991; Suarez-Villa and Cuadrado Roura 1993).

What sets Austria apart from most other national experiences is a very low level of local developmental effects in the hinterland, coupled with substantial transborder transactions that pre-empted any local externalisation of functions. At the same time, a seemingly paradoxical emphasis on R&D allowed for competitiveness in targeting export market niches, mainly through the clustering of research-intensive establishments in the capital region. This has resulted in a dualistic structure that has its roots in both the international-spatial and the intra-corporate divisions of labour. As will be shown in the following sections, this phenomenon has resulted in a significantly different organisational orientation towards research and innovation that is spatially grounded through localised transactional and co-operative arrangements.

3   Operational Objectives and Organisation in Austria's Electronics Industry

The analysis of organisational characteristics relied on a postal and interview-based survey of manufacturing establishments from all Austrian regions. The sample of 58 establishments (14% of the total of 415 electronics manufacturing plants) was geographically representative, with 40% of all respondents located in the capital region, compared with 41% for the actual total population (and 60% of respondents from the hinterland regions, compared with the actual population's proportion of 59%). The largest proportion of respondents was in the industrial (42%) and component (38%) electronics sectors. Consumer electronics was therefore under-represented in the sample, although chi-squared tests did not reveal any significant biases. Austria's electronics industry favoured industrial and component goods production over the past 20 years, and the sample's composition reflects the strong emphasis on these two sectors (see Table 3). Most of the establishments included in the sample were domestic owned (70%), with average establishment size closely following the actual population's estimates without any significant bias. In the absence of other relevant information for the actual population of establishments, there are no indications that the sample was biased in terms of plant age, ownership, production characteristics or input and product data.

Table 3   Composition of sample

	% of establishments[a]
Location	
Capital region	39.7
Hinterland	60.3
Corporate status	
Single establishment	43.9
Multiestablishment organisation:	56.1
Main plant	22.8
Branch plant	33.3
Total employment	
8 – 49	39.3
50 – 99	17.9
100 – 499	32.1
≥ 500	10.7
Sector[b]	
Industrial electronics	42.3
Component electronics	38.5
Consumer electronics	19.2

[a] based on a national sample of 58 establishments, authors' survey, March 1992; [b] SIC sectoral categories: industrial electronics (SIC 3571, 3572, 3575); component electronics (SIC 3671, 3672, 3674, 3675, 3676, 3677, 3678); consumer electronics (SIC 3651); classified according to each establishment's principal product.

A survey of operational motivations, shown in Table 4, reveals some significant differences between the hinterland's and the capital region's establishments. Most important is the weighting of patenting and innovation licensing amongst the capital region's establishments. This reflects a greater emphasis on R&D in the capital region's plants, where 29% of respondents declared patenting and innovation licensing to be their most important operational motivation, as opposed to 13% in the hinterland. Whilst this is consistent with an export-intensive, niche market-driven strategy, particularly in the industrial and component electronics sectors, it further emphasises the dualistic structure of Austria's electronics industry and its spatial division of labour. Innovation is at the core of the differences between hinterland and capital region, in so far as establishment operation is concerned. In the hinterland, therefore, inward investment has not only prevented the local externalisation of enterprise functions, it has also a diminished emphasis on product innovation.

Table 4   Primary motivations for electronics plant operation (%)

	Capital Region[a]	All other regions
Organisational interest or experience	41.2	32.3
Patent or license on a product	29.4	12.9
Potential market niche	11.8	19.4
Linkage with foreign firms	11.8	9.7
Government subsidies	5.9	12.9
Profits	0.0	12.9

[a] includes the Vienna, Burgenland and Lower Austria regions. All estimates are based on data obtained from the authors' survey, 1992 ($N = 48$).

Patenting is usually one of the most significant outcomes of R&D, given its potential to safeguard proprietary rights, particularly in product-embodied innovations (Mohnen et al. 1986; Tödtling 1987, 1990; Fischer and Menschik 1994; Fischer et al. 1994; Pornschlegel 1992). It is also one of the most risk-prone aspects of R&D, since the eventual outcomes or economic success of most invention projects are very difficult to anticipate (Mansfield 1971; Griliches 1990; Suarez-Villa 1990, 1993; Ács and Audretsch 1991). The uncertainty of patenting as a proprietary right is further heightened whenever its benefits are embodied in a product and cross international boundaries, where ownership rights may be reduced by imitation or through relaxed enforcement (see, for example, Scherer 1983; Swann 1986; Fischer et al. 1994). Despite these difficulties, it is obvious that patenting is an important operational priority, and may actually be less uncertain than any other alternative in a competitive, export-oriented industry targeting specific market niches.

Patenting also seems to be an essential component of incremental innovation, since the resulting advances tend to be more firm specific and are more narrowly targeted to improve predetermined attributes. The importance of patenting for incremental innovation therefore lies in the competitive character of Austria's electronics exports, since the pre-emption of an innovation by any other firm would entail either significant disadvantages and delays in marketing goods, or substantial additional costs in obtaining a licensing arrangement from an actual or potential competitor. From a locational perspective, the benefits of patenting for incremental innovation are reinforced by concentration, where cumulative infrastructural and human resource investments can develop a critical mass of externalities (Suarez-Villa and Hasnath 1993). Concentrating in the capital region can also provide opportunities to save or generate resources for innovation, by allowing greater specialisation in R&D-intensive productive operations whilst externalising others through subcontracting.

The establishment of market niches is relatively important for the hinterland's plants, with 19% of all sampled firms declaring this criterion their prime operational motivation (see Table 4). This was the second highest-ranking choice for the hinterland's establishments, and it reflects the importance of sustaining export market niches, as part of the hinterland branch plants' strategic 'fit' in the intra-corporate and international divisions of labour. In contrast with the capital region's establishments, however, the development of market niches for the hinterland's plants occurs primarily through lower cost advantages and their position in the intra-corporate division of labour, rather than through R&D intensity or innovation. This once again underscores the structural cleavage between the hinterland's and the capital region's operations, where a dualistic division of labour based on innovative capabilities has driven the spatial distribution of electronics production.

Considerations related to managerial self-development, learning and self-promotion obviously played an important part in the choice of the most important criterion, organisational interest and experience, by both hinterland and capital region establishments. The ranking of this motivation seems to be consistent with the rise of a managerial business system in Austria's electronics industry, where decisions made by salaried managers and top executives have tended to reflect more individual concerns and agendas (see, for example, Donaldson and Lorsch 1986; Illeris 1986; Chandler 1990; Eisenhardt and Schoonhoven 1990; Nelson 1991). Individual perceptions of organisational challenges, such as overcoming difficulties in production technology, in labour relations, in co-ordinating suppliers or in charting marketing strategies, appear to be important incentives for risk-taking with respect to this criterion. The greater importance of this factor amongst the capital region's establishments, chosen by 41% as their most important motivation, reflects its higher priority among R&D-intensive firms, where finding creative solutions to many organisational problems tends to be a more pressing component of daily management.

A more creative organisational interest would seem to be important for managing the sort of co-operative arrangements that externalisation can foster. It would also be important for targeting an organisation's priorities in a fluid market environment, and for devising the means to achieve the firm's goals, amongst which the saving of resources to reinvest in R&D would be a prime objective. Managing internal flexibility as, for example, in shifting labour and personnel between production and R&D to facilitate the rapid introduction of product innovations, requires more creative managerial interest. The significantly higher ranking of this motivation amongst the capital region's establishments (41%, versus 32% for the hinterland plants in Table 4) also reflects the structural cleavage in production between hinterland and capital region. The spatial division of labour for electronics production therefore encompasses Austria's managerial business system, where the externalisation of functions achieved through concentration has apparently conditioned the individual expectations of the managerial class.

The majority presence of branch plants, which are linked from the start with foreign parent plants or contractors, was the likely cause of the hinterland establishments' lower rating of the foreign linkage criterion. Intra-corporate transactions may therefore account for a lower appreciation of foreign linkages as a primary motivator among hinterland plants, since for all purposes this is a *de facto* component of their strategic division of labour. Linkages with foreign concerns appear, on the other hand, to be a greater priority for the capital region's R&D-intensive establishments, which tend to be more on their own, use more domestic capital and rely more on independent transactions with foreign suppliers and marketing concerns. The higher rating of government subsidies for the hinterland establishments can be attributed mainly to local incentives related to tax abatement and the provision of land. The relative insignificance of this criterion shows the limited impact that public incentives have as an operational motivator. Such incentives, limited as they may be, have nevertheless been considered important by some locales as they attempt to build up a cluster of firms and to expand their employment base, however modestly (see, for example, Stöhr 1990; Moore et al. 1991).

A higher priority for longer-term concerns related to innovation and market share may be the root cause of the lacking in importance of profits as a primary motivation amongst the capital region's establishments (see Table 4). Profits are known to be more of a shorter-term concern in manufacturing organisations, particularly amongst export-intensive firms marketing technologically advanced goods (see, for example, Mueller 1986; Suarez-Villa and Han 1991). Although low-ranking, profits were relatively more important amongst hinterland establishments, reflecting the concerns of branch plant operators and their position within intra-corporate divisions of labour. In such organisations, operational viability may be more closely linked with short-term earnings, where continuous cost reduction is often an overarching concern. Indeed, a branch operation's successful 'fit' within the broader corporate division of labour may be largely determined by its short-term performance on profits and costs. In general, however, the insignificance of profits as a primary motivation should be a source of reflection about a central tenet of economic theory, where profit-seeking has been traditionally regarded as the most important cause of business action.

To assess the organisational characteristics of Austria's electronics plants, a factor analysis of 13 production indicators, obtained from the sample of establishments, was undertaken (see Table 5). Seven metric and six categorical variables were included in the analysis, providing information on aspects such as process technology, labour, skills, capital, and input and product transactions. Indicators with categorical terms became necessary whenever obtaining metric data posed obstacles, due to record-keeping difficulties by the establishments, or because of privacy concerns. The factor analysis will provide an analytical platform upon which statistical analyses of establishment characteristics related to R&D, sub-contracting, production technology and location can be developed.

Table 5   Variables included in the factor analysis

Variable name	Definition
Age	Year of reference: 1992
Sector	Industrial electronics (0), component and consumer electronics (1)
Plant type	Branch (0), single plant (1)
Production	Mass production schedule (0), small batch schedule (1)
Ownership[a]	Foreign (0) or domestic (1); over 50% of total capital
Employment size	Total employment per plant
R&D personnel ratio	R&D personnel per total employment ratio
Skills ratio	Skilled production labour per semi-skilled or unskilled worker
Labour ratio	Non-production personnel per total production labour
Labour costs[a]	Payroll outlays per employee
Gender ratio	Male personnel per female employed
Domestic inputs[a]	Input purchases from domestic suppliers: Less than 50% (0), 50% or more (1)
Domestic sales[a]	Product sales to domestic buyers: Less than 50% (0), 50% or more (1)

[a] in 1992 Austrian Schillings

The factor analysis of organisational indicators, shown in Table 6, reveals five main factors, three of which (process, scale, skills) are most significant (with eigenvalues of 1 or greater). The results obtained with the process factor provide support for the importance of process technology over scale, as a major influence in the organisation of electronics production (see, for example, Hickson et al. 1969; Aldrich 1979; Marsh and Mannari 1981; Tushman and Anderson 1986; Barley 1990). How technology and work were organised, in a process sense, appear to be more important than any given single production indicator, such as labour cost, skills, sales or capital. Organisational capability underlies organisational performance in a general sense, and seems to be consistent with the revealed preference of organisational interest and experience as the most important operational motivation, as previously discussed. How organisations are structured, in a strategic sense, influences the internal and external projection of the firm, in aspects such as the interaction between production and R&D, in the externalisation of activities through co-operative arrangements and, less directly, in its eventual market performance (see, for example, Tushman and Anderson 1986; Garvin 1988; Hansen 1988; Lazonick 1990; Whittaker 1990; Gillespie 1991; Chandler 1992; Meyer and Scott 1992; Morroni 1992).

The results obtained with the process factor show that electronics plants with small batch production schedules tend to employ fewer R&D personnel relative to total employment, are more likely to have stronger domestic transactions and tend to be more typical of the industrial electronics sector. It should be noted, however, that the relationship between small batch production and R&D employment does not necessarily reflect a lower R&D intensity for such processes. Establishments utilising small batch production processes tend to be smaller in size and therefore have a lower proportion of R&D employment, compared with establishments that employ more labour or produce more output. Their R&D activities, therefore, have a more limited operational scope, with R&D resources being more concentrated on a shorter range of tasks, attempting to achieve greater effectiveness. The stronger domestic product sales generally reflect greater interplant subcontracting and domestic producer interrelations for plants with small batch production routines. Both subcontracting and small batch production schedules are more likely to be found in the industrial electronics sector, where production in limited runs is more frequent than in the component or consumer electronics sectors.

Table 6   Factor analysis of organisational characteristics in Austria's electronics establishments[a]

Process		*Scale*		*Skills*		*Linkages*		*Capital*	
Production	0.799	Employment size	.869	Skills ratio	0.718	Domestic inputs	0.885	Ownership	0.681
R&D personnel ratio	-0.786	Plant type	745	Domestic sales	-0.664				
Sector	-0.692	Labour costs	.597						
Domestic sales	0.580								
Eigenvalue	3.164		045		1.433		0.987		0.833
Cumulative variance	0.243		401		0.511		0.587		0.651

[a]   varimax rotation (only factor loadings greater than 0.500 are shown). $N = 56$

The second most significant factor, scale, shows that smaller establishments tend to be single plant operations and are more likely to have lower labour costs. The cost effectiveness of specialised small producers appears to be significant for a research-intensive industry pursuing niche markets, with limited capital and strong international competition. These findings are consistent with the results obtained for other industries and nations, where smaller-size establishments have been found to be lower-cost independent producers (Fischer and Nijkamp 1988; Dougherty 1989; Bosworth et al. 1990; Suarez-Villa and Han 1990b; Loveman and Sengenberger 1991; Pratten 1991; Carlsson 1992; Miller 1992). Smaller establishments are also more likely to be independent producers, and their cost effectiveness appears to be greater than those of branch operations. In emphasising larger scale and establishment 'fit' within a given corporate division of labour, branch plants may therefore be missing the significant cost advantages associated with smaller, independent operations. Nevertheless, for most multilocational organisations, co-ordinating a large network of small, independent producers may be far too complex an alternative to branch plant operation and investment.

   The results obtained with the skills factor indicate that plants employing more skilled labour tend to export a higher proportion of their production. More and better-skilled labour may therefore be more important for Austria's international competitiveness in electronics, over and above other considerations, such as its labour cost advantage with respect to Germany. This finding supports the results of earlier research in other nations, where a direct link has been found between higher proportions of skilled labour and stronger firm performance (see, for example, Bartel and Lichtenberg 1987; Wozniak 1987; Lazonick 1990; Suarez-Villa and Han 1990a; Gillespie 1991; Hansen 1991). Greater utilisation of skilled labour would also be important for research-intensive establishments, adding to internal flexibility in shifting personnel from R&D to production (or vice versa), and in the rapid introduction of product innovations. As will be shown in the following section, labour skills and organisational size are an important influence in the allocation of resources for R&D, in the externalisation of functions and in Austria's overall spatial division of labour for electronics production.

4   Territorial Concentration and Export-Driven R&D

Much of Austria's international advantage in electronics manufacturing, and in successfully developing innovative product market niches, has resulted from the research-intensive capabilities of its firms and their ability to position themselves favourably within the spatial division of labour. Such favourable positioning obviously relies on the availability of skills, the benefits of adjusting organisational size and the types of production processes utilised. Whilst the previous analyses have shown the significance of process, scale and skills in Austria's electronics production, little is known about the association of those factors with R&D. Theoretically, it can be expected that significant spatial differences in R&D intensity may be found between the capital region's establishments, for example, and those that locate in more isolated areas or in smaller places in the hinterland (see, for example, Tödtling 1984, 1987; Fischer and Menschik 1991, 1994).

Most hinterland establishments are either foreign- or domestic-owned branch plants that have limited R&D capabilities and local transactions. Higher R&D intensity may well be a function of concentration in a large industrial area, such as the capital region, where better access to labour skills and to opportunities to adjust establishment size through externalisation can be found (see, for example, Howells 1984; Davelaar and Nijkamp 1989; Davelaar 1991; Alderman and Fischer 1992; Kleinknecht and Poot 1992). Thus, whilst a strong expansion of exports promoted the territorial dispersion of electronics production through branch establishments, the very basis for the longer-term competitiveness of the industry may lie in the primate industrial concentration of the capital region.

A set of analyses was undertaken to determine the statistical association between the most representative variables of each organisational factor (process, scale, skills), and the R&D intensity of the sampled production establishments. Geographical differentiation was introduced to test for any possible contrasts between the capital region and the hinterland by including two test sets: one with the sampled establishments spatially undifferentiated; and another with only the hinterland plants. The statistical tests include separate regressions between each of the organisational factor variables and the sampled establishments' level of R&D outlays (as the dependent variable). The results of the analyses, shown in Table 7, provide insights on the hinterland-capital region differentiation of R&D intensity, the significance of spatial concentration to support it and the relationship between organisational size, skilled labour and R&D resource allocation.

Significant contrasts between capital region and hinterland can be found in the association of R&D outlays with two of the organisational variables (employment size, skilled production labour) tested. The effect of a capital region location can be observed in the results obtained with the aggregate data (all regions' estimates, Table 7) and those obtained using only the hinterland data. The latter are statistically insignificant and have a low association: in contrast, when the capital region's establishments are included (upper portion, Table 7), the estimates provide a strong association and are statistically significant at the 1% confidence level. In general, the externalities generated through concentration are important for research-intensive establishments, where better access to skilled labour is a prerequisite to maintain the kind of internal flexibility needed for rapid innovation and product quality improvements. Concentration also seems to help firms adjust their organisational size to fit their operational requirements, where external flexibility is influenced by the variety of transactional possibilities that are available.

The stronger effect of labour skills on R&D outlays may be explained by the incremental and continuous character of innovation in many research-intensive establishments, where practical learning (learning-by-doing) tends to occur jointly between production and R&D. Such establishments are more likely to value labour skills, to the extent that the internal flexibility they create helps the rapid introduction of product innovations. The stronger effect of labour skills over plant size should provide for some re-evaluation of previous research on the link between R&D and plant or firm size (see, for example, Mowery 1983; Cuneo and Mairesse 1984; Pakes and Schankerman 1984; Pavitt et al. 1987; Ács and Audretsch 1991; Kleinknecht et al. 1991). Larger establishment size, by and of itself, does not seem to be the most important determinant of R&D. A larger, more skilled labour force is more conducive to greater R&D intensity in electronics production. It can help implement product and process innovations faster, while improving quality control. Highly skilled labour can often play a dual role in production and R&D by alternatively shifting into and out of R&D activities and production tasks as needed. R&D and higher labour skills may therefore enjoy a synergistic relationship that is greatly determined by spatial concentration (Felsenstein and Shachar 1988; Cohen and Levinthal 1989; Pornschlegel 1992).

The results obtained with the process variable (production schedule) appear invariant between hinterland and capital region. The effect of introducing the capital region's establishments is therefore insignificant, indicating that a capital region-hinterland differentiation has little effect on the association between production schedules and R&D (see Table 7). In general, small batch production schedules tend to have a weaker relationship with R&D, indicating that processes with higher output volumes are more likely to have a stronger R&D function (see, for example, Pakes and Schankerman 1984; Mohnen et al. 1986; Pisano 1990; Pornschlegel 1992). This reflects the importance of the association between a larger, more skilled labour force and R&D, as discussed earlier; larger output can be expected of establishments that employ more skilled labour and such organisations are also more likely to be research-intensive. This is consistent with the results of the factor analysis, where the production variable (in the process factor) provided an inverse relationship between small batch production and R&D employment (see Table 6).

Table 7   Organisational factors and R&D in Austria's electronics establishments

Dependent variable		Production schedule (process)	Employment size (scale)	Skilled production labour (skills)
		Independent variables		
All regions				
R&D outlays	(a)	85.200 (21.932)	−9.422 (8.132)	−1.415 (6.590)
	(b)	−81.635 (24.199)	0.105 (0.016)	0.262 (0.034)
	R^2	0.304	0.614	0.700
	F	11.381**	41.277**	60.560**
Hinterland (all regions except Vienna, Burgenland, Lower Austria)				
R&D outlays	(a)	31.000 (11.205)	4.702 (7.465)	7.646 (6.649)
	(b)	−28.273 (12.641)	0.015 (0.016)	0.022 (0.049)
	R^2	0.294	0.065	0.016
	F	5.003*	0.831	0.194

* significant at the 5% level; ** significant at the 1% level or better; standard errors are shown in parentheses; (a): constant, (b): OLS regression coefficient

It should be noted that the sample data utilised above was evenly balanced between the hinterland and the capital region. One-half of all the sampled plants declaring any R&D outlays were located in the hinterland regions. The majority of establishments declaring R&D outlays were in the industrial electronics sector, with production process characteristics that were quite similar across establishments. Most of the sampled establishments (77%) declaring any R&D outlays were domestic owned, and most (68%) located their R&D operations at their headquarters. Of all establishments declaring any R&D outlays, 43% exported over 50% of their production directly; by comparison, the similar proportion of output exported directly by capital region plants (with or without any R&D outlays) was 33%, versus 62% in the hinterland. This seems to be consistent with the characteristics of branch plants operating in the hinterland areas of other countries, where low local embeddedness tends to be the norm (see, for example, Hargrave 1985; Grabher 1993; Phelps 1993; Turok 1993a). Concentration in the capital region therefore reduces the direct export linkages of plants with R&D

operations, due to the higher level of localised transactions, which include both hierarchical and co-operative arrangements.

It may be expected that a higher level of local transactions amongst the capital region plants may lead to lower shipping expenditures. Concentration may lead to a reduced frequency of long-distance shipping, as local transactions account for a higher proportion of inputs and subcontracting, effectively replacing the more expensive, longer-range contacts. In the hinterland, on the other hand, the lack of embeddedness and local alternatives may be expected to promote farther reaching transactions, with higher shipping expenditures. This should result in a stronger statistical association between the level of establishment R&D outlays and shipping expenditures in the hinterland. The lack of embeddedness and the territorial dispersion of hinterland plants could thus potentially reduce the availability of resources that can be allocated to R&D in the hinterland, as the organisational stock of resources is diminished by higher shipping costs. In some respects, therefore, the perception of lower production costs associated with hinterland locations may be rather deceptive, given the longer-term opportunity costs that may be incurred by limiting investment in research and innovation:

$$\text{Freight costs} = 1.991 + 0.206 \quad \text{R\&D outlays} \tag{1}$$
$$\quad\quad (1.038) \quad\quad (0.052)$$
$$R^2 = 0.441 \quad F = 15.755$$
$$\text{(all regions)}$$

$$\text{Freight costs} = 1.714 + 0.216 \quad \text{R\&D outlays} \tag{2}$$
$$\quad\quad (1.485) \quad\quad (0.018)$$
$$R^2 = 0.948 \quad F = 144.432$$
$$\text{(hinterland)}$$

Equations (1) and (2), above, show contrasting results for the statistical association between establishments' freight costs and their R&D expenditures. Equation (1) includes data for all of the Austrian region's sampled electronics plants, and results in a lower association (0.441) than Equation (2) (0.948) which includes only the hinterland plants. The results for both tests were statistically significant at better than the 1% level of confidence. Thus, the introduction of the capital region's establishments in Equation (1) causes the strength of the association to be more than halved. This is consistent with the expectation of a higher level of local transactions for the capital region plants, as concentration provides greater possibilities for subcontracting and other transactions (see, for example, Williamson 1985; Imrie 1986; Lorenz 1988; Scott and Kwok 1989; Antonelli 1992; Thoburn and Takashima 1992).

It should be noted that larger circumstances and influences, beyond the localised benefits of externalisation, had an effect on the capital region's concentration of production establishments. For many years, Eastern Europe's borders were practically closed to transborder production transactions, thereby limiting the possibilities to expand the contact networks of electronics producers. Vienna's proximity to the closed Eastern borders may have, therefore, had a confining effect on electronics production, increasing the localised transaction intensity of the capital region. At the same time, Vienna's more distant location from West Germany and other Western European nations, compared with most of the hinterland, may have also helped foster the localised network of productive transactions through higher shipping and communication costs.

5   R&D, Subcontracting and Just-in-Time

This section extends the previous analyses to consider the relationship between R&D, subcontracting and the adoption of just-in-time (JIT) production techniques in Austria's electronics industry. The majority of establishments engaged in subcontracting are in the capital region, indicating the importance of concentration for locally externalised operations, whether of a hierarchical or a relational-co-operative character. In Austria's case, 67% of all sampled establishments sub-contracting out part of their production were engaged with producers and suppliers located within the same urban region. Furthermore, 85% of establishments subcontracted by other firms were located within 20 km of the contracting plants.

Locating nearby is obviously important for subcontracting in Austria's electronics industry, as it allows better possibilities to implement innovations and adjustments in production more quickly. A nearby location can provide better access to a subcontractor's operation, possibly allowing familiarisation with processes and procedures that can increase personal trust and promote mutually helpful arrangements. Proximity to other potential subcontractors within a metropolitan region can also provide alternatives if performance becomes unsatisfactory and, at the same time, it can be an incentive for subcontracted firms to improve quality and efficiency. Even where subcontracting has involved primarily hierarchical or 'arm's-length' transactions, proximity helps communication between firms, particularly whenever contracts require renegotiation or delivery deadlines can not be met.

In Austria's capital region, localised subcontracting has provided a way for firms to gain access to a more finely graded division of labour, by allowing opportunities to externalise or 'network' production tasks. Subcontracting seems to allow greater flexibility in production and in saving firms' resources for crucial internal operations, such as R&D, in the face of requirements for continuous and rapid technological change and all the uncertainties associated with it. In Austria's capital region, some firms have been able to downsize or fragment themselves into greater specialisation, thereby targeting their R&D efforts more specifically towards certain products. This has also allowed them to limit in-house functions or component needs. In some cases, the operations subcontracted out to other firms duplicate the internal routines of the contractors, allowing more flexibility to deal with changes in demand, and with the need to differentiate or customise some products rapidly through R&D-derived efforts.

Subcontracting basically involves two types of transactions. One is inter-dependent, functioning as a co-operative or relational mode, with flexible bound-aries and long-term horizons. The second is more of a contractual, competitive or hierarchical (or 'arm's-length') mode, where interfirm trust and familiarity may be very limited, or missing altogether. The first type in various ways expects sub-contractors to add value beyond the simple transactional requirements, such as providing knowledge and expertise on the development of the goods they supply, co-ordinating design and quality control with the contractors' own production routines, and having the willingness to co-ordinate or reduce output whenever market demand subsides, regardless of initial expectations (see, for example, Imrie 1986; Scott and Kwok 1989; Barley 1990; Turok 1993a).

In Austria's case, it is virtually impossible to determine precisely to what extent one or the other type of transaction is prevalent amongst establishments. The lack of published data makes it necessary to rely on establishment surveys, where information on the specifics of any ongoing or previous transactions tends to be incomplete, at best. It seems, however, that both types of transactions are common and tend to coexist, particularly amongst capital region establishments, often being entered into simultaneously with separate firms. This reveals some of the com-plexity of subcontracting, where a firm may dualistically engage in both types of transactions, and where the determination to engage in one or the other form may hinge on previous interfirm experiences, perceived reputation, imitation of competitors' arrangements or even managerial personalities and friendships. As will be discussed later, this complexity in contractual transactions is reflected in the prevalence of *two-way* subcontracting, whereby firms that subcontract out part of their production are also contracted by other firms. What seems certain in the Austrian case is that the concentration of electronics establishments in the capital region is very important for subcontracting, and that such arrangements have a substantial effect on firms' capabilities to engage in R&D.

Besides providing opportunities for subcontracting, concentration can promote the local diffusion of R&D knowledge, shared informally through such means as the local hiring of experienced personnel, of consultants with knowledge of com-petitors' operations, through social contact between firms' R&D personnel, or through the kinds of manufacturers' conferences that metropolitan regions with a critical mass of similar establishments can support on a regular basis. Concen-tration therefore provides a peculiar coexistence of co-operation and competition, where the co-operative transactions realised through subcontracting can end up supporting the most competitive component of the firm, R&D, through resource savings derived from increasing specialisation, and by avoiding the implemen-tation of costly production techniques, such as JIT, whenever externalisation can be an effective substitute. Resources saved by subcontracting can be reinvested in R&D, increasing its effectiveness by hiring new personnel or purchasing equipment, by co-ordinating research activities more closely with production to speed up the introduction of innovations, or by linking up with international research laboratory networks. Austria's competitive emphasis on R&D-based

incremental innovation is therefore supported by subcontractual transactions, where innovation tends to be more firm specific and is more narrowly focused on product development.

Research activities typically tend to be the most important internal source of long-term competitiveness for advanced technology firms. As such, they are also the most private or secretive components of the firm, and are the ones least likely to be shared with, or entrusted to, outsiders. Research is also the least likely area to be externalised, at least in any major way, since subcontractual relations can be quite subject to instability, depending on market demand fluctuations, supplier quality and timely performance, or matters of personal trust. Most firms therefore tend to view R&D as an asset to be primarily deployed for competitive purposes, where rapid technological change and the maintenance or expansion of market niches is a major concern. A recent study of UK producer services by O'Farrell et al. (1993), for example, found that the percentage of manufacturing establishments using external suppliers for R&D declined substantially with increasing firm size, with the steepest decline occurring in firms within the 50-100 employment range. In Austria's case, there is little evidence that any R&D operations are being subcontracted out to any significant extent, domestically or internationally.

The complexity of the interrelations involved in subcontracting production is reflected by the fact that the majority (56%) of establishments subcontracting out were also being contracted by others. At the same time, a large proportion (62%) of the plants that were contracted by others was also subcontracting out part of their own production. *Two-way* subcontracting was more relevant for firms with complex production processes, such as those in industrial electronics (telecommunications equipment, computers, industrial and professional instruments), where downsizing and specialisation could contribute to greater efficiency. Firms pursuing market niche objectives could benefit more from this strategy, given the frequent need to adjust production runs and product specifications to forestall competitive challenges. Niche market specialisation also appeared to make such firms attractive as subcontractors, for tasks and products for which they possessed qualified personnel and facilities, but which were not directly related to their specific market objectives. Two-way subcontracting therefore allowed some establishments to reduce idle capacity during demand fluctuations, or to gain the greater specialisation required by a market niche orientation. Two-way subcontracting also appeared to reduce overhead costs, providing savings on resources that could then be deployed to support R&D activities. In general, the narrower specialisation favoured by market niche production was supportive of two-way subcontracting, especially when establishments' technological capabilities exceeded the narrower scope of their production objectives.

Two-way subcontracting may be part of a trust-building strategy if it limits the number of partners and achieves mutually advantageous co-production arrangements. As such, it may result in a fairly closed network of establishments, where collective trust is built up over time through frequent interaction. In such arrangements, the kind of single sourcing typically assumed of co-operative transactions would be superseded by a larger sense of community amongst the participants, where multiple sourcing within the network would be the norm. Such networked co-operation might possibly lead to consortia for production, allowing its participant establishments to develop larger projects or complex products that one or two firms would not be able to handle on their own. Research operations may not necessarily be included, or even shared amongst participants, with R&D remainning a highly private internal function, despite the fact that all firms would stand to benefit from the innovations that are marketed.

Despite the potential advantages noted above, two-way subcontracting can nevertheless result in a very different scenario, where trust is ultimately diminished if an establishment moves towards a diversified relationship with several firms, some of which may be actual or potential competitors. In such cases, a breakdown of trust might be avoided when diversification occurs with the acceptance of all participants. Otherwise, two-way subcontracting may well occur at the future expense of interfirm trust, or at its periphery, with adverse effects for the firms that engage in it without the tacit or explicit consent of their partners. It is obvious, therefore, that two-way subcontracting tends to defy the notion of single sourcing or of exclusive bilateral transactions as the premier expression of trust.

It is virtually impossible to determine precisely if two-way subcontracting is occurring more through trust than at its expense in Austria's electronics industry, but the evidence available through the survey of establishments indicates that it is not a significant problem for maintaining co-operative relations in general. Indeed, the high proportion of establishments engaged in two-way subcontracting indicates that it may well be a trust-building arrangement rather than a conflict-prone adaptation. Also, two-way subcontracting occurs in establishments engaging in co-operative or in hierarchical–competitive transactions, reflecting its bimodal suitability in the process of externalisation. In virtually all cases, however, two-way subcontracting does not involve R&D activities, indicating the internal and very private way in which firms regard this function.

If two-way subcontracting indeed contributes to increase trust, the externalised (or subcontracted) and the internalised (or in-house) contracted operations may be largely complementary rather than potential substitutes. As internal production capabilities grow qualitatively, requests from other firms can be accommodated, even though the contracted establishment may already be subcontracting out part of its own production. These simultaneous and seemingly paradoxical processes of internalisation and externalisation have, for example, been found to occur in UK producer services by O'Farrell et al. (1993), and in business service firms by Marshall (1989), where the net balance tended towards vertical integration in both manufacturing and business services. In Austria's case, two-way subcontracting does not appear to be a source of vertical reintegration, although it is impossible to determine precisely whether, on balance, this phenomenon has led more towards the internalisation of functions than their externalisation.

The seemingly paradoxical coexistence of internalisation and externalisation in two-way subcontracting reveals a transactional complexity that was largely unanticipated by Coase's (1937) pioneering work on industrial organisation. Assuming industrial organisation to be governed by the evolving interface between internal and external transactive interactions, Coase assumed that firms would internalise any additional task until its cost equalled or surpassed that of externalising it. The qualitative and external relational value of these two dimensions was lost in Coase's strict cost accounting of internalisation, however. It appears that, in two-way subcontracting, firms tend to overexternalise, reinternalising then some functions by accepting contracts from other firms, as a way to build up co-operative relations that can be helpful in times of stress or of higher market demand for their own products. Far from being a systematic sort of miscalculation or perceptive bias, overexternalisation seems to be part of a deliberate strategy aimed at accommodating other firms' needs in expectation of reciprocity or support during difficult periods.

The reinternalisation of functions through two-way subcontracting may also occur involuntarily through, for example, idle capacity in times of low market demand, prompting firms to seek additional work in order to maintain employment and short-term revenues. Sustaining employment and skilled labour is particularly important for Austria's electronics firms, since jobs tend to be fairly permanent over the long term and often carry implicit security guarantees. Furthermore, occasional shortages of skilled personnel in some occupations make it necessary to retain as many in-house labour skills as possible. The internalisation of functions in two-way subcontracting can also occur whenever there is a learning advantage in being contracted by firms developing new processes or significant product innovations where, for example, producing a new component for a contractor can raise the innovative standards of the contracted firm. Partner firms experiencing higher than expected market demand may also influence a subcontracted firm's decision to reinternalise some tasks, at least temporarily, as a way to ensure reciprocity from its partner organisations.

Only a small proportion (26%) of the plants engaged in subcontracting implemented just-in-time production methods, indicating a possible trade-off between subcontracting and the JIT adoption decision. All indications are that subcontracting was used to avoid the costs of adopting JIT methods, especially in the industrial electronics sector (where limited production runs are more typical) with greater uncertainty, and which can differ significantly from one batch to another. For a market niche-oriented firm, the adoption of JIT technology can involve substantial expenses, related to programming, technical supervision, shop floor management, supply co-ordination and the modification of facilities. JIT production methods are both more expensive and more difficult to implement for limited production runs, or for product lines that require frequent redesign and innovation to remain competitive. Such expenditures, increased by frequent modifications and adjustments, typically draw resources away from other important components of the firm, such as R&D and marketing. Greater production flexibility could therefore be obtained through subcontracting than through the adoption of JIT methods, freeing up resources that could be deployed to other areas of the firm (see, for example, Ahmed et al. 1991; Linge 1991).

Another aspect limiting the adoption of JIT production technology is the emphasis on incremental and continuous product innovation. Continuous innovation requires greater flexibility in production, and the ability to seek expertise externally, where subcontracting production out can be a most important means of access. Continuous innovation is less compatible with standardised, high volume production, where JIT methods would be more effective. With continuous innovation, it is often difficult to anticipate when a product will undergo significant differentiation, requiring changes in components and other supplies, and rendering some of the previously adopted production routines obsolete. The emphasis therefore tends to be more on structuring production in limited runs, or customising it to a significant extent. Nevertheless, although continuous innovation seems to be a factor limiting the adoption of JIT, the most significant constraint is the very limited availability of 'patient', long-term resources from financial institutions and the public sector to support the costs of conversion.

Despite the stronger local and domestic transactions that subcontracting involves, many of the establishments engaged in this activity export a significant amount of their production on their own. Almost two-thirds of the sampled plants, subcontracting out part of their production, export over 45% of their output directly. Similarly, one-half of the plants employing JIT production methods export over 60% of their output directly. The stronger local transactions that subcontracting and JIT involve therefore tend to complement and enhance, rather than reduce, the export intensity of production. In general, however, hinterland plants export more of their production directly, but the direct export performance of subcontracting or JIT establishments, most of which are in the capital region, nevertheless remains high.

Domestic capital has obviously played an important role in subcontracting production or applying JIT methods. Most of the sampled plants subcontracting (70%) or using JIT methods (69%) were domestic owned. R&D activities were also important in domestic-owned establishments that subcontracted production or that used JIT methods. The vast majority of the plants engaged in subcontracting production (95%) or applying JIT methods (89%) had R&D operations. A significant emphasis on R&D by domestic-owned establishments therefore complemented the export intensity and implementation of either subcontracting or JIT methods. Because of the very limited application of JIT methods in plants subcontracting production, however, it is obvious that domestic capital tended to favour subcontracting over JIT production whenever the latter could be avoided or substituted by the former. This further confirms the existence of a significant tradeoff between these two possibilities, as a way to redeploy scarce resources towards research and innovation.

Statistical tests of the organisational variables' association with R&D only for establishments subcontracting production or utilising JIT methods (Table 8) shows significant differences with the earlier results provided in Table 7. The new results reveal a much stronger association between R&D and the variables representing scale and skills (employment size, skilled production labour) when only the subcontracting establishments are considered. A stronger association is also obtained when only establishments that were simultaneously subcontracting out part of their production, and that were being subcontracted by other firms, are included. Similarly, the associations between R&D and the scale and skill factor variables are much stronger for plants utilising JIT production methods. All of the new tests including these variables also have a higher statistical significance than those previously shown in Table 7.

Table 8   R&D, subcontracting and just-in-time production in Austria's electronics establishments

Dependent variable		Production schedule (*process*)	Employment size (*scale*)	Skilled production labour (*skills*)
			Independent variables	
Plants subcontracting out				
R&D outlays	(a)	130.333 (41.495)	−13.765 (5.463)	−1.579 (9.439)
	(b)	−125.896 (48.658)	0.161 (0.009)	0.336 (0.035)
	R^2	0.427	0.974	0.912
	F	6.694*	332.955**	93.747**
Plants subcontracted and subcontracting out				
R&D outlays	(a)	153.000 (73.549)	−17.169 (7.730)	−12.881 (14.056)
	(b)	−149.375 (90.079)	0.164 (0.010)	0.345 (0.038)
	R^2	0.407	0.986	0.953
	F	2.750	284.282**	80.946**
Plants using JIT				
R&D outlays	(a)	105.000 (47.586)	−25.029 (12.027)	7.158 (8.051)
	(b)	−98.500 (67.297)	0.166 (0.016)	0.340 (0.026)
	R^2	0.263	0.946	0.967
	F	2.142	105.416**	173.910**

* significant at the 5% level; ** significant at the 1% level or better; standard errors are shown in parentheses; (a): constant, (b): OLS regression coefficient

Comparing the results provided in Table 7 and Table 8, the stronger association obtained for R&D and scale (employment size) in establishments subcontracting or using JIT methods (Table 8), indicate greater R&D intensity with increasing plant size. The elasticities obtained with the new results also favour establishments subcontracting (those subcontracting out, and those simultaneously subcontracting and subcontracted, Table 8) or using JIT methods. Thus, increases in size are more likely to lead to greater R&D outlays in establishments subcontracting or using JIT technology. On the latter, however, it should be recalled that the proportion of establishments using JIT production methods was relatively unimportant; the results of Table 8 therefore hold greater significance for plants engaged in subcontracting, which encompass the majority of electronics producers, as previously discussed. Plants that simultaneously engaged in subcontracting out and that were subcontracted by other firms provided the strongest association, indicating a slightly higher advantage over operations that were only subcontracting out. Subcontracting thus appears to be a preferred means to reduce organisational entropy through downsizing, and to economise on resources that can potentially be reinvested in R&D. Larger-size establishments are in a better position to do this, due to their usually larger resource base, and because they have better possibilities of deploying the managerial expertise required by subcontracting (see, for example, Williamson 1985; Chandler 1990; Antonelli 1992; Thoburn and Takashima 1992).

The application of JIT production methods can also contribute to a reallocation of resources towards R&D by reducing inefficiencies. The implementation of JIT methods that, for example, produce cost savings by reducing inventory and overhead capital, or that help improve quality control, can release resources for other needs. (O'Grady 1988; Ahmed et al. 1991). Larger-size establishments tend to be in a better position to achieve such a reallocation of resources, particularly when they have longer production runs or more standardised product segments. Larger-size establishments are also in a better position to maintain a broader portfolio of R&D projects if resources can be saved and reallocated to the latter. A larger portfolio of R&D projects can also serve as a risk-reducing strategy in the larger establishments, especially when research is more basic (or less narrow and firm specific) and its potential success is more uncertain.

The association between R&D and the labour skills variable also improved for both subcontracting and JIT establishments, as shown in Table 8. Comparing the results of the individual regressions between R&D outlays and the labour skills and plant-size variables, in Table 7 and Table 8, it can be observed that labour skills provided a higher elasticity than plant size. Larger R&D outlays can therefore be expected in plants employing more skilled production labour, given similar sizes. This further confirms the importance of human capital resources, over that of size or scale, for Austria's innovation and export-driven electronics industry (see, for example, Bartel and Lichtenberg 1987; Wozniak 1987; Bosworth et al. 1990; Lazonick 1990).

The association between R&D and the labour skills variable was also strong in plants utilising JIT production methods, as shown in Table 8. The application of JIT production technology requires higher production skills, given the need for greater co-ordination, individual and group involvement, quality control and precise production programming. Such operations were more likely to have an R&D component, given the importance of production labour skills for R&D. A stronger association between R&D and labour skills in subcontracting plants can also be found, comparing the results of Table 7 and Table 8. Subcontracting allowed establishments to shift highly skilled personnel to support R&D, through downsizing or specialisation, or to labour in production tasks that benefited more directly from R&D operations. Larger R&D outlays were, therefore, also more likely to be found in subcontracting organisations where greater specialisation and a more effective use of scarce physical and human capital resources could occur.

The strengths provided by higher labour skills and human capital resources, in general, were a most important asset in maintaining internal flexibility in most of Austria's electronics establishments. Highly skilled labour is more likely to have greater versatility, it is more apt to be used interchangeably amongst the various facets of production and it can advance faster through the learning process needed to implement product and process innovations. Higher labour skills are also more likely to contribute to R&D, since research projects often require preliminary production testing, where labour's insights and experience are essential to reduce flaws before manufacturing starts. In research units, skilled labour often also performs important tasks, assisting in the set-up of experiments and in the development of new ideas from the start, where first-hand knowledge of production can greatly influence product design and workmanship.

6   Conclusion

The progressive territorial distribution experienced by Austria's electronics industry over the past 20 years has resulted in a dualistic productive structure, where research-intensive establishments have tended to remain concentrated in the capital region, whilst poorly embedded branch operations have proliferated in the hinterland. Given the advantages of concentration for R&D, Austria's capital region may well determine the future competitiveness of electronics manufacturing. Concentration has, in general, provided better opportunities for localised externalisation and better access to skills that have helped generate the kinds of resource savings and tradeoffs that favour R&D.

Reallocating scarce resources to bolster R&D has been important for domestic-owned firms serving export market niches. Austria's electronics establishments have generally favoured subcontracting over the adoption of just-in-time methods, thereby downsizing or achieving greater specialisation whilst generating more resources for R&D. The complexity of subcontracting and its coexistence with the internalisation of some activities was reflected in the high incidence of two-way subcontracting. Engaging in two-way subcontracting therefore meant that the skilled personnel and facilities available could also be applied to work in tasks contracted for by other firms, as a way to elicit co-operation or to reduce idle productive capacity. The fact that two-way subcontracting, and subcontracting in general, mostly occurred amongst establishments located nearby, further underscores the advantages of concentration for research-intensive firms.

The importance of skilled production labour in research-intensive establishments underscores the significance of human capital resources for market niche-oriented firms. The synergism between R&D and production labour skills tends to be most advantageous in the rapid introduction of product innovations, where the versatile allocation of personnel is an important asset. Such versatility adds to the internal flexibility of firms in areas related to R&D, and was important for plants engaged in subcontracting or applying JIT methods. Concentration, therefore, provided access to a more finely detailed division of labour where externalisation promoted greater specialisation, internal flexibility and expanded research capabilities.

The significance of concentration for research and innovation in Austria's electronics industry has important policy implications for the development of advanced technology industries. Developing R&D capabilities may be the most important single source of long-term technological competitiveness in advanced countries. Thus, support for concentration and opportunities for the localised externalisation of functions may perhaps be the most important contribution that policy action can make towards developing an effective industrial strategy. This presents a thorny dilemma for public action, since it is rather the dispersion or progressive territorial distribution of industrial activity that can have an impact on employment and social equity. Thus, policy action aimed at fostering and expanding technological competitiveness may well produce results at the expense of greater territorial disparities for high-technology employment, human capital development and labour income.

The bimodal and seemingly paradoxical coexistence of co-operation and competition observed with respect to subcontracting and the internal proprietary emphasis on R&D can therefore be extended to the policy arena. Policy programmes aimed at achieving balance may therefore have to adopt a bimodal approach, encouraging concentration for research-intensive firms whilst promoting dispersion for branch operations, as the most feasible way to increase both hinterland employment and technological competitiveness. Such a strategy would perhaps be the only means for policy programmes to function effectively within a dualistic productive structure. It is noteworthy, however, that, in the Austrian case, changes in the spatial division of labour for electronics production have occurred with so little specific policy intervention. Austria's spatial division of labour in electronics production pretty much developed on the basis of opportunistic corporate initiatives, as German and Western European restructuring deepened, and Austrian firms sought competitive market niches in the West. It therefore remains to be seen if, in other countries, policy action can overturn the kind of structural dualism that a strong export orientation, based on continuous research and innovation on the one hand and on branch plant production on the other, has generated in the Austrian case.

Acknowledgements: Klaus Schuch, research assistant at the Vienna University of Economics and Business Administration, translated and distributed survey questionnaires, conducted selected interviews and compiled part of the primary and secondary data for this research. His assistance with these tasks is much appreciated. Dianne Christianson's efforts in processing the manuscript and tables are gratefully acknowledged. Research support received from Vienna University of Economics and Business Administration is also acknowledged and appreciated.

References

Ács Z. and Audretsch D.B. (1991): R&D, firm size, and innovative activity. In: Ács Z. and Audretsch D.B. (eds.) *Innovation and Technological Change: An International Comparison*, Harvester Wheatsheaf, New York, pp. 39-59

Ahmed N.U., Tunc E.A. and Montagno R.V. (1991): A comparative study of US manufacturing firms at various stages of just-in-time, *Production Research* 29, 787-802

Alderman N. and Fischer M.M. (1992): Innovation and technological change: An Austrian-British comparison, *Environment and Planning A* 24 (2), 273-288

Aldrich H. (1979): *Organisations and Environments*. Prentice-Hall, Englewood Cliffs [NJ]

Altmann N., Kohler C. and Meil P. (eds.) (1992): *Technology and Work in German Industry*, Routledge, London

Amin A. and Malmberg A. (1992): Competing structural and institutional influences on the geography of production in Europe, *Environment and Planning A* 24 (3), 401-416

Amin A., Charles D.R. and Howells J. (1992): Corporate restructuring and cohesion in the new Europe, *Regional Studies* 26 (4), 319-32

Amsden A.J. (1990): *Asia's Next Giant*, Oxford University Press, Oxford [UK]

Antonelli C. (1992): *The Economics of Information Networks*, North-Holland, Amsterdam

Austria Offert (1991): Das neue Europa, *Austria Offert* 17, 4-41

Aydalot P. (1984): Reversals of spatial trends in French industry since 1974. In: Lambooy J.G. (ed.) *New Spatial Dynamics and Economic Crisis*, Finnpublishers, Tampere, pp. 41-63

Bade F.J. and Kunzmann K.R. (1991): Deindustrialization and regional development in the Federal Republic of Germany. In: Rodwin L. and Sazanami H. (eds.) *Industrial Change and Regional Economic Transformation: The Experience of Western Europe*, Harper Collins Academic, London, pp. 70-104

Barley S.R. (1990): The alignment of technology and structure through roles and networks, *Administrative Science Quarterly* 35 (1), 61-103

Bartel A.P. and Lichtenberg F.R. (1987): The comparative advantage of educated workers in implementing new technology, *The Review of Economics and Statistics* 69 (1), 1-11

Bayer K. (1992): Austria: The impact of '1992' on the Austrian manufacturing industries. In: European Free Trade Association (ed.) *European Economic Integration: Effects of '1992' on the Manufacturing Industries of the EFTA Countries*, EFTA, Vienna, pp. 59-113

Becattini G. (1989): Sectors and districts: Some remarks on the conceptual foundations of industrial economics. In: Goodman E., Bamford J. and Saynor P. (eds.) *Small Firms and Industrial Districts in Italy*, Routledge, London, pp. 123-135

Best M.H. (1990): *The New Competition: Institutions of Industrial Restructuring*, Harvard University Press, Cambridge [MA]

Bosworth D.L., Jacobs C. and Lewis J.A. (1990): *New Technologies, Shared Facilities and the Innovatory Firm*, Avebury, Aldershot

Braun E. and Macdonald S. (1982): *Revolution in Miniature: The History and Impact of Semiconductor Electronics*, Cambridge University Press, Cambridge [MA]

Braun P. and Polt W. (1988): High technology and competitiveness: an Austrian perspective. In: Freeman C. and Lundvall B.-Å. (eds.) *Small Countries Facing the Technological Revolution*, Pinter, London, pp. 203-225

Brusco S. (1990): The idea of the industrial district: its genesis. In: Pyke F., Becattini G. and Sengenberger W. (eds.) *Industrial Districts and Inter-firm Co-operation in Italy*, International Institute for Labour Studies, Geneva, pp. 10-19

Buckley P.J. and Ghauri P.N. (eds.) (1983): *The Internationalization of the Firm*, Academic Press, London

Bundeskammer der Gewerblichen Wirtschaft (1990): Forschung und Dokumentation in Österreich 1989, Bundeskammer der Gewerblichen Wirtschaft, Vienna

Cantwell J. (ed.) (1992): *Multinational Investment in Modern Europe: Strategic Interaction in the Integrated Community*, Edward Elgar, Aldershot [UK], Brookfield [VT]

Cappellin R. and Batey P.W.J. (eds.) (1993): *Regional Networks, Border Regions and European Integration*, Pion, London

Cappellin R. and Nijkamp P. (eds.) (1990): *The Spatial Context of Technological Development*, Avebury, Aldershot

Carlsson B. (1992): The rise of small business: Causes and consequences. In: Adams W.J. (ed.) *Singular Europe: Economy and Policy of the European Community after 1992*, University of Michigan Press, Ann Arbor, pp. 145-170

Chandler A.D. (1992): Organisational capabilities and the economic history of the industrial enterprise, *Journal of Economic Perspectives* 6 (3), 79-100

Chandler A.D. (1990): *Scale and Scope: The Dynamics of Industrial Capitalism*, Harvard University Press, Cambridge [MA]

Cheshire P.C. (1991): Problems of regional transformation and deindustrialization in the European Community. In: Rodwin L. and Sazanami H. (eds.) *Industrial Change and Regional Economic Transformation: The Experience of Western Europe*, Harper Collins Academic, London, pp. 237-267

Coase R. (1937): The nature of the firm, *Economica* 4 (16), 386-405

Cohen W.M. and Levinthal D.A. (1989): Innovation and learning: The two faces of R&D, *Economic Journal* 99 (397), 569-596

Cuneo P. and Mairesse J. (1984): Productivity and R&D at the firm level in French manufacturing industry. In: Griliches Z. (ed.) *R&D, Patents, and Productivity*, University of Chicago Press, Chicago, pp. 375-392

Davelaar E.J. (1991): *Regional Economic Analysis of Innovation and Incubation*, Avebury, Aldershot

Davelaar E.J. and Nijkamp P. (1989): The role of metropolitan milieu as an incubation centre for innovations: A Dutch case study, *Urban Studies* 26, 517-525

Donaldson G. and Lorsch J.W. (1986): *Decision-Making at the Top: The Shaping of Strategic Decisions*, Basic Books, New York

Dorfman N.S. (1987): *Innovation and Market Structure: Lessons from the Computer and Semiconductor Industries*, Harper and Row, Cambridge [MA]

Dosi G., Pavitt K. and Soete L. (1988): *The Economics of Technical Change and International Trade*, Harvester Wheatsheaf, Brighton

Dougherty D.C. (1989): *Strategic Organisation Planning: Downsizing for Survival*, Quorum, New York

Dunning J.H. (1988): The UK's international direct investment Position in the mid-1980s. In: Dunning J.H. (ed.) *Multinationals, Technology and Competitiveness*, Unwin Hyman, London, pp. 327-347

Eisenhardt K.M. and Schoonhoven C.B. (1990): Organisational growth: Linking founding team, strategy, environment, and growth among US semiconductor ventures, 1978-1988, *Administrative Science Quarterly* 35 (3), 504-529

Ernst D. (1983): *The Global Race in Microelectronics*, Campus, Frankfurt

Fachverband der Elektro- und Elektronikindustrie (1974-92): Elektro- und Elektronik-industrie: Statistischer Bericht. Fachverband der Elektro- und Elektronikindustrie, Vienna

Felsenstein D. and Shachar A. (1988): Locational and organisational determinants of R&D employment in high technology firms, *Regional Studies* 22 (6), 477-486

Fischer M.M. (1990): The micro-electronic revolution and its impact on labour and employment. In: Cappellin R. and Nijkamp P. (eds.) *The Spatial Context of Technological Development*, Avebury, Aldershot, pp. 43-74

Fischer M.M. and Menschik G. (1994): *Innovationsaktivitäten in der österreichischen Industrie. Eine empirische Untersuchung des betrieblichen Innovationsverhaltens in ausgewählten Branchen*, Abhandlungen zur Geographie und Regionalforschung 3, University of Vienna, Vienna

Fischer M.M. and Menschik G. (1991): Innovation und technologischen Wandel in Österreich, *Mitteilungen der Österreichischen Geographischen Gesellschaft* 133, 43-68

Fischer M.M. and Nijkamp P. (eds.) (1987): *Regional Labour Markets: Analytical Contributions and Cross-National Comparisons*, North-Holland, Amsterdam

Fischer M.M. and Nijkamp P. (1988): The role of small firms in regional revitalization, *The Annals of Regional Science* 18, 28-42

Fischer M.M., Fröhlich J. and Gassler H. (1994): An exploration into the determinants of patent activities: Some empirical evidence for Austria, *Regional Studies* 28 (1), 1-12

Fischer M.M. and Schuch K. (1994): Die Österreichische Zulieferindustrie im Lichte des Europäischen Integrationsprozesses, WSG-RR 5, Department of Economic and Social Geography, Vienna University of Economics and Business Administration

Freeman C. (1974): *The Economics of Industrial Innovation*, Penguin Books, Harmondsworth [UK]

Fritsch M. (1992): Regional differences in new firm formation: Evidence from West Germany, *Regional Studies* 26 (3), 233-242

Garvin D.A. (1988): *Managing Quality*, Free Press, New York

Gillespie R. (1991): *Manufacturing Knowledge*, Cambridge University Press, Cambridge [MA]

Grabher G. (ed.) (1993): *The Embedded Firm: On the Socioeconomics of Industrial Networks*, Routledge, London

Griliches Z. (1990): Patent statistics as economic indicators: A survey, *The Journal of Economic Literature* 28 (4), 1661-1707

Håkansson H. (1989): *Corporate Technological Behaviour: Co-operation and Networks*, Routledge, London

Hansen N.M. (1992): Competition, trust, and reciprocity in die development of innovative regional milieux, *Papers in Regional Science* 71, 95-106

Hansen N.M. (1991): Factories in Danish fields: How high-wage flexible production has succeeded in peripheral Jutland, *International Regional Science Review* 14, 109-132

Hansen N.M. (1988): Regional consequences of structural changes in the national and international division of labor, *International Regional Science Review* 11, 105-119

Hargrave A. (1985): *Silicon Glen: Reality or Illusion? Global View of High Technology in Scotland*, Mainstream Publishing, Edinburgh

Henderson J. (1991): *The Globalisation of High Technology Production: Society, Space and Semiconductors in the Restructuring of the Modern World*, Routledge, London

Hickson D.J., Pugh D.S. and Pheysey D.C. (1969): Operations technology and organisation structure: An empirical reappraisal, *Administrative Science Quarterly* 14 (3), 378-397

Hounshell D.A. and Smith J.K. (1988): *Science and Corporate Strategy*, Cambridge University Press, Cambridge [MA]

Howell T.R., Noellert W.A., Maclaughlin J.H. and Wolff A.W. (1987): *The Microelectronics Race*, Westview, Boulder

Howells J.R.L. (1984): The location of research and development: Some observations and evidence from Britain, *Regional Studies* 18 (1), 13-29

Hughes K. (1986): *Exports and Technology*, Cambridge University Press, Cambridge [MA]

Illeris S. (1986): New firm creation in Denmark: The importance of the cultural background. In: Keeble D. and Wever E. (eds.) *New Firms and Regional Development in Europe*, Croom Helm, London, pp. 141-150

Imrie R.F. (1986): Work decentralisation from large to small firms: A preliminary analysis of subcontracting, *Environment and Planning A* 18 (7), 949-966

Jacquemin A. (1987): *The New Industrial Organisation: Market Forces and Strategic Behavior*, MIT Press, Cambridge [MA]

Jacquemin A. and Sapir A. (eds.) (1989): *The European Internal Market: Trade and Competition*, Oxford University Press, Oxford [UK]

Jessop R., Nielsen K., Kastendiek H. and Pederson O.K. (eds.) (1991): *The Politics of Flexibility: Restructuring State and Industry in Britain, Germany and Scandinavia*, Edward Elgar, Aldershot [UK], Brookfield [VT]

Jones G. and Schröter H.G. (eds.) (1992): *The Rise of Multinationals in Continental Europe*, Edward Elgar, Aldershot [UK], Brookfield [VT]

Keeble D., Owens P.L. and Thompson C. (1983): The urban-rural manufacturing shift in the European Community, *Urban Studies* 20, 405-418

Kleinknecht A. and Poot T. P. (1992): Do region matter for R&D? *Regional Studies* 26 (3), 221-232

Kleinknecht A., Foot T. P. and Reijnen J.O.N. (1991): Formal and informal R&D and firm size. In: Ács Z. and Audretsch D.B. (eds.) *Innovation and Technological Change: An International Comparison*, Harvester Wheatsheaf, New York, pp. 94-108

Krugman P. (1991): *Geography and Trade*, MIT Press, Cambridge, [MA]

Kubin I. and Steiner M. (1992): Labor market performance and regional types: A conceptual framework with empirical analysis of Austria, *International Regional Science Review* 14, 275-298

Lane C. (1989): *Management and Labour in Europe: The Industrial Enterprise in Germany, Britain and France*, Edward Elgar, Aldershot [UK], Brookfield [VT]

Lawton Smith H. (1991): Advanced technology industry in Oxfordshire: location of markets and competitors, *Urban Studies* 28, 205-218

Lazonick W. (1990): *Competitive Advantage on the Shop Floor*, Harvard University Press, Cambridge [MA]

Linge G.J.R. (1991): Just-in-time: More or less flexible? *Economic Geography* 67 (4), 316-332

Lorenz E. (1988): Neither friends nor strangers: Informal networks of subcontracting in French industry. In: Gambetta D. (ed.) *Trust: Making and Breaking Co-operative Relations*, Basil Blackwell, Oxford, pp. 194-210

Loveman G. and Sengenberger W. (1991): The re-emergence of small-scale production, *Small Business Economics* 3, 1-37

Maier G. and Tödtling F. (1986): Towards a spatial deconcentration of entrepreneurial control? Some empirical evidence for the Austrian regions, 1973-1981, *Environment and Planning A* 18 (1), 209-224

Maier G. and Weiss P. (1986): The importance of regional factors in the determination of earnings: The case of Austria, *International Regional Science Review* 10, 211-220

Malecki E.J. (1991): *Technology and Economic Development. The Dynamics of Local, Regional and National Change*, Longman, Harlow, Essex and Wiley, New York

Mansfield E. (1971): *Research and Innovation in the Modern Corporation*, W.W. Norton, New York

Marschak T.A., Glennan T.K. and Summers R. (1967): *Strategy for R&D*, Springer, Berlin, Heidelberg, New York

Marsh R.M. and Mannari H. (1981): Technology and size as determinants of the organisational structure of Japanese factories, *Administrative Science Quarterly* 26 (1), 33-57

Marshall J.N. (1989): Corporate reorganization of the geography of services: Evidence from the motor vehicle aftermarket in the West Midlands Region of the UK, *Regional Studies* 23 (2), 139-150

Mason C., Pinch S. and Witt S. (1991): Industrial change in Southern England: A case study of the electronics and electrical engineering industry in the Southampton city-region, *Environment and Planning A* 23 (5), 677-703

Mathias P. and Davis J.A. (eds.) (1991): *Innovation and Technology in Europe: From the Eighteenth Century to the Present*, Blackwell, Oxford [UK], Cambridge [MA]

Matis H. (ed.) (1994): *The Economic Development of Austria since 1870*, Edward Elgar, Aldershot [UK], Brookfield [VT]

Mattsson L.-G. and Stymne B. (eds.) (1991): *Corporate and Industry Strategies for Europe: Adaptations to the Single Market in a Global Industrial Environment*, North-Holland, Amsterdam

Matzner E. and Wagner M. (1990): *The Employment Impact of New Technology: The Case of West Germany*, Avebury, Aldershot

McCalman J. (1989): *The Electronics Industry in Britain*, Routledge, London

Meyer J.W. and Scott R. (1992): *Organisational Environments*, Sage, Newbury Park [CA]

Miles I. (1988): *Home Informatics: Information Technology and the Transformation of Everyday Life*, Pinter, London

Miller G.J. (1992): *Managerial Dilemmas: The Political Economy of Hierarchy*, Cambridge University Press, Cambridge [MA]

Milne S. (1990): New forms of manufacturing and their spatial implications: the UK electronic consumer goods industry, *Environment and Planning A* 22 (2), 211-232

Mohnen P., Nadiri M. I. and Prucha I. (1986): R&D, production structure, and rates of return in the US, Japanese and German manufacturing sectors, *European Economic Review* 30 (4), 749-771

Molle W. and Klaassen L.H. (1985): *Industrial Mobility and Migration in the European Community*, Gower, Aldershot

Moore B., Tyler P. and Elliott D. (1991): The influence of regional development incentives and infrastructure on the location of small and medium-sized companies in Europe, *Urban Studies* 28 (1), 1-26

Morroni M. (1992): *Production Process and Technical Change*, Cambridge University Press, Cambridge [MA]

Mowery D.C. (1983): Industrial research and firm size, survival, and growth in American manufacturing, 1921-1946: An assessment, *Journal of Economic History* 43, 953-980

Mueller D.C. (1986): *Profits in the Long Run*, Cambridge University Press, Cambridge [MA]

Nelson R.R. (1991): Why firms differ, and how does it matter? *Strategic Management Journal* 12, 61-74

Oakey R.P., Thwaites A.T. and Nash P.A. (1980): The regional distribution of innovative manufacturing establishments in Britain, *Regional Studies* 14, 235-253

O'Grady P.J. (1988): *Putting Just-in-time Philosophy into Practice*, Kogan Page, London

O'Farrell P.N., Moffat L.A.R. and Hitchens D.M.W.N. (1993): Manufacturing demand for business services in a core and peripheral region: Does flexible production imply vertical disintegration of business services? *Regional Studies* 27 (5), 385-400

Österreichisches Institut für Wirtschaftsforschung (1991): Statistische Übersichten. WIFO-Monatsberichte, Vienna

Österreichisches Statistisches Zentralamt (1969-89): Österreichische Industriestatistik, Österreichisches Statistisches Zentralamt, Vienna

Pakes A. and Schankerman M. (1984): An exploration into the determinants of research intensity. In: Griliches Z. (ed.) *R&D, Patents, and Productivity*, University of Chicago Press, Chicago, pp. 209-232

Pavitt K. (1986): Chips and 'trajectories': How will the semiconductor influence the sources and directions of technical change? In: Macleod R. (ed.) *Technology and the Human Prospect*, Pinter, London, pp. 31-54

Pavitt K., Robson M. and Townsend J. (1987): The size distribution of innovative firms in the UK: 1945-1983, *Journal of Industrial Economics* 35, 291-316

Phelps N.A. (1993): Branch plants and the evolving spatial division of labour: A study of material linkage change in the Northern region of England, *Regional Studies* 27 (2), 87-101

Piore M.J. and Sabel C. (1984): *The Second Industrial Divide*, Basic Books, New York

Pisano G.P. (1990): The R&D boundaries of the firm: An empirical analysis, *Administrative Science Quarterly* 35 (1), 153-176

Polt W. (1992): Technology programmes in small open economies: A review of recent experiences, Discussion Paper OEFZS-4662, Austrian Research Centers Seibersdorf, Austria

Pornschlegel H. (ed.) (1992): *Research and Development in Work and Technology – Proceedings of a European Workshop*, Dortmund, Germany, 23-25 October 1990, Physica-Verlag, Heidelberg

Porter M.E. (1990): *The Competitive Advantage of Nations*, Free Press, New York

Pratten C. (1991): *The Competitiveness of Small Firms*, Cambridge University Press, Cambridge [MA]

Quévit M. (1992): The regional impact of the internal market: A comparative analysis of traditional industrial regions and lagging regions, *Regional Studies* 26 (4), 349-360

Ratti R. (1988): Development theory, technological change and Europe's frontier-regions. In: Aydalot P. and Keeble D. (eds.) *High Technology Industry and Innovative Environments: The European Experience*, Routledge, London, pp. 197-220

Sabel C. (1989): Flexible specialisation and the re-emergence of regional economies. In: Hirst P. and Zeitlin J. (eds.) *Reversing Industrial Decline?* Berg Publishers, Oxford, pp. 17-70

Schackmann-Fallis K.P. (1989): External control and regional development within the Federal Republic of Germany, *International Regional Science Review* 12, 245-261

Scherer F.M. (1983): The propensity to patent, *International Journal of Industrial Organisation* 1, 221-225

Scott A.J. (1988): *New Industrial Spaces*, Pion, London

Scott A.J. and Kwok E.C. (1989): Inter-firm subcontracting and locational agglomeration: A case study of the printed circuits industry in Southern California, *Regional Studies* 23 (5), 405-416

Sheppard E., Maier G. and Tödtling F. (1990): The geography of organisational control: Austria, 1973-1983, *Economic Geography* 66 (1), 1-21

Soete L. and Dosi G. (1983): *Technology and Employment in the Electronics Industry*, Pinter, London

Steiner M. and Sturn D. (1992): From coexistence to co-operation: The consequences of open borders in Styria, Paper presented at the 32nd European Congress of the Regional Science Association, Louvain-la-Neuve

Stöhr W. (1990): *Global Challenge and Local Response: Local Initiatives for Economic Regeneration in Contemporary Europe*, Mansell, London

Stöhr W. (1986): Regional innovation complexes, *Papers of the Regional Science Association* 59, 29-44

Suarez-Villa L. (1993): The dynamics of regional invention and innovation: Innovative capacity and regional change in the twentieth century, *Geographical Analysis* 25 (2), 147-164

Suarez-Villa L. (1990): Invention, inventive learning and innovative capacity, *Behavioral Science* 35, 290-310

Suarez-Villa L. and Cuadrado-Roura J. R. (1993): Thirty years of Spanish regional change: Interregional dynamics and sectoral transformation, *International Regional Science Review* 15, 121-156

Suarez-Villa L. and Han P.-H. (1991): Organisations, space and capital in the development of Korea's electronics industry, *Regional Studies* 25 (4), 327-343

Suarez-Villa L. and Han P.-H. (1990a:) International trends in electronics manufacturing and the strategy of industrialization, *Rivista Internazionale di Scienze Economiche e Commerciali* 37, 381-407

Suarez-Villa L. and Han P.-H. (1990b): The rise of Korea's electronics industry: Technological change, growth and territorial distribution, *Economic Geography* 66 (3), 273-292

Suarez-Villa L. and Hasnath S. (1993): The effect of infrastructure on invention: Innovative capacity and the dynamics of public construction investment, *Technological Forecasting & Social Change* 44, 333-358

Surholt V. (1984): Die Entwicklung der Elektroindustrie in Österreich nach dem zweiten Weltkrieg, Doctoral Thesis, Vienna University of Economics and Business Administration

Swann G.M.P. (1986): *Quality Innovation: An Economic Analysis of Rapid Improvements in Microelectronic Components*, Pinter, London

Teece D.J. (1980): Economies of scope and the scope of the enterprise, *Journal of Economic Behavior & Organization* 1 (3), 223-247

Thoburn J.T. and Takashima M. (1992): *Industrial Subcontracting in the UK and Japan*, Avebury, Aldershot

Thwaites A.T. and Alderman N. (1989): *Industrial Research and Development in the Regional Economy*, Belhaven Press, London

Todd D. (1989): *The World Electronics Industry*, Routledge, London

Tödtling F. (1990): Regional differences and determinants of entrepreneurial innovation: Empirical results from an Austrian case study. In: Ciciotti N., Alderman N. and Thwaites A.T. (eds.) *Technological Change in a Spatial Context*, Springer, Berlin, Heidelberg, New York, pp. 259-284

Tödtling F. (1987): The regional pattern of industrial R&D in Austria: Sectoral, organisational and locational determinants, *Revue d'Economie Regionale et Urbaine* 2, 239-254

Tödtling F. (1984): Organisational characteristics of plants in core and peripheral regions of Austria, *Regional Studies* 18 (5), 397-412

Turok I. (1993a): Inward investment and local linkages: How deeply embedded is Silicon Glen? *Regional Studies* 27 (5), 401-417

Turok I. (1993b): Contrasts in ownership and development: Local versus global in Silicon Glen, *Urban Studies* 30 (2), 365-386

Tuseman M.L. and Anderson P. (1986): Technological discontinuities and organisational environments, *Administrative Science Quarterly* 31, 439-465

Veltz P. (1991): New models of production organisation and trends in spatial development. In: Benko G. and Dunford M. (eds.) *Industrial Change and Regional Development*, Belhaven Press, London, pp. 193-204

Watts H. (1981): *The Branch Plant Economy: A Study of External Control*, Longman, London, New York

Whittaker D.H. (1990): *Managing Innovation: A Study of British and Japanese Factories*, Cambridge University Press, Cambridge [MA]

Williamson O.E. (1985): *The Economic Institutions of Capitalism: Firms, Markets, and Relational Contracting*, Free Press, New York

Wozniak G.D. (1987): Human capital, information, and the early adoption of new technology, *Journal of Human Resources* 22 (1), 101-112

4 Information Processing, Technological Progress, and Retail Market Dynamics

with *J. Cukrowski*

The chapter analyses the potential impact of technological progress in information processing on the size of retail markets. The analysis – restricted to a single commodity market with uncertain demand – shows that the ability of firms to process information and predict demand may affect the characteristics of retail markets. The results, moreover, indicate that risk-averse firms always devote resources to demand forecasting, producers are better off trading with retailers than with final consumers, and the volume of output supplied through retail markets is greater than it would be if producers traded directly with consumers.

1 Introduction

Corporations are generating information at an exponential rate. It is three forces that are fueling the rate of growth: the use of informative technology to automate business processes, the implementation of enterprise resource planning systems, and the diffusion of information technology within the business. As the quantity of information collected and stored by corporations has soared, managers have realised that information is an enterprise's most valuable asset. It contains a record of past corporate performance. A clear vision of the past, present and future is essential to sustain competitivity in the rapidly changing business world. The dynamic application of information often separates the success from the failures in many sectors of the economy. The interfaces between developments in information technology, retailing strategies and consumer behaviour are attracting an increasing amount of attention in the marketing and economic literature. Most of this literature either summerises recent developments or speculates about future developments and their economic and social impacts (see e.g., Webster 1994; Jeannet and Hennessey 1995; De Canio and Watkins 1998). This present contribution takes a somewhat different stance. The view is taken that technological progress in information processing increases the number of firms operating in retail markets.

The existence of retail markets and the role of retail firms are traditionally explained by spatial factors and increasing returns to scale in transportation, storage or in the acquisition and dissemination of information about the quality, range, and prices of products available (see, e.g., Heal 1980; Wilson 1975). In many cases, such as retail trade in services or in goods which can not be transported or stored, most of these factors are irrelevant. The major shortcoming of the earlier literature on retail markets is that it attempts to view retail trade only from the supply side, and the models used have no explicit reference to demand, especially to demand uncertainty, which is natural in most markets. This current contribution makes a modest attempt to show that retail trade (or at least part of it) may not be connected with economies of scale and that it can be explained exclusively by the rational behaviour of firms operating in a stochastic environment. Since retailers are, by definition, intermediaries between consumers and suppliers, they can serve as a buffer between suppliers and a market with demand uncertainty, and, in particular, they can bear the risk associated with demand fluctuations. Thus, to analyse retailing convincingly one needs to explicitly model the interaction between consumers and suppliers in an uncertain environment. Only this type of model can credibly explain why there is a need for retailers as middlemen and what determines the characteristics of the retail markets.

As has been already recognised in the literature, in real life firms are never sure about a number of variables such as factor prices, the exact shape of the production function or the market demand curve. Even if firms are certain of their cost structures, they very rarely (if ever) know precisely which demand conditions they face. The behaviour of firms operating in markets with uncertain demand has been analysed in several studies (see Sandmo 1971; Leland 1972; Lim 1980). However, in most of these papers, the firms' beliefs about demand are summarised in a subjective probability distribution which can not be changed by the firms' actions. The fact that the firm may be able to predict changes in demand, or at least to decrease the range of possible variations, is usually neglected in the standard studies of economic behaviour under uncertainty. Nevertheless, the ability of the firm to predict demand, although not always perfect, may affect a number of parameters of economic equilibrium (see, e.g., Nelson 1960, for an analysis of uncertainty and prediction in competitive markets). The conjecture in this paper is that market analysis, information processing and demand forecasting activities not only affect the characteristics of economic equilibrium when the producer sells goods directly to the final consumers, but also influence the equilibrium characteristics of retail markets.

In the analysis which follows we assume that the relationship between quantities demanded and market prices randomly varies from period to period, and that demand analysis is both costly and time consuming. In particular, we focus on the market where the total demand originates from a large (but finite) number of sources. The demand curve in each individual source changes randomly from period to period, but in any time period demand changes are assumed to be correlated with the changes prior to this period, reflecting a certain inertness in consumer behaviour. Since information gathering and processing requires time, the sum of individual demands (i.e., the total demand) can not be instantaneously determined. In particular, we assume that the results of the market analysis are available only after the end of the period. Consequently, the firms' output-price decisions have to be made based not on the current demand but on its prediction. In each period the profit maximising firms set their volumes of output, since it has a high commitment value within a period of time (i.e., the output decisions are irreversible within the time unit). The price is assumed to be more flexible and can change to some extent due to real market conditions. However, firms operating in the market are still assumed to be unable to learn the true demand function during the period of time, and, consequently, have to rely only on the results of the demand analysis. Since demand forecasts are based on past data, a prediction error appears, and, consequently, firms' output decisions always deviate from what is optimal.

To focus solely on the role of uncertainty and data processing (to avoid the problem of inventories, transportation and storage), one can think about a supplier of services, such as a sightseeing tour operator, operating in the market with demand depending on the weather in a season. We suppose that the supplier can analyse the market in order to decrease the variance of demand fluctuations. Moreover, the supplier is assumed to set capacity before the season (i.e., before real demand becomes known) and has two options: to sell services directly to final consumers during the season (at an uncertain price), or to sell services forward (before the season) to retail firms (at a fixed price, lower, however, than the expected price to consumers). In such a set-up, we examine how technological progress in information processing can affect the size of retail markets.

The paper is organised as follows. Market demand is characterised in Section 2. Section 3 shows how forecasts of actual demand can be computed. Section 4 provides an analysis of the optimal demand forecasting strategy in the monopolistic supplier and the retail firms. In Section 5 the alternative methods of distribution of the output produced (i.e., with and without retail firms) are considered. The implication of technological progress in data processing on the size of retail markets is presented in Section 6. The concluding section summerises some of the major findings of the study.

2   Uncertain Demand

Consider a market where total demand originates from a large number of identical sources N (one can think of these sources as consumers). Suppose that demand in each individual source i ($i = 1, 2, ..., N$) at any period of time t (t is an integer number, $-\infty < t < +\infty$) is linear with an additive random term $\eta_{i,t}$ (for the sake of simplicity, assume that random variables $\eta_{i,t}$ are identically distributed with zero mean and finite variance ($\sigma_{i,t}^2 = \sigma_i^2$). Total inverse demand at period t is

$$P_t(Q_t, \eta_{1,t}, \eta_{2,t}, ..., \eta_{N,t}) = a - b\,Q_t + \sum_{i=1}^{N} \eta_{i,t} \tag{1}$$

where $Q_t = \Sigma_{i=1}^N q_{i,t}$ is the total quantity demanded at price $P_t = (P_t \geq 0)$, a and b are positive constants.

The random variables $\eta_{i,t}$ can move up or down in response to changes in the variables omitted from a correct demand specification, such as, for instance, interest rates, inflation, personal income, prices of other goods, etc. Much of this movement, however, might be due to factors which are hard to capture, such as, for example, changes in the weather or in consumer tastes. Thus, in many cases it may be difficult (or even impossible) to explain fluctuations in demand through the use of a structural model. Moreover, it might happen that, even if statistically significant regression equations can be estimated, the result will not be useful for forecasting purposes (for example, when explanatory variables which are not lagged must themselves be forecasted). In such situations, an alternative means of obtaining predictions of $\eta_{i,t}$ has to be used. The easiest way is to predict changes in $\eta_{i,t}$ based on the analysis of their movements in the past. Such forecasts, however, are possible only if the random variables $\eta_{i,t}$ are observable and if they are correlated with their previous values.

3   Demand Forecasting

To simplify the analysis, assume that random deviations $\eta_{i,t}$ ($i = 1, 2, ..., N$) from the expected values of individual demands are independent[1] and described by identical stationary stochastic processes with a memory (e.g., by autoregressive processes of any order).[2] In other words, assume that for any individual demand, variances and covariances of random variables, $\eta_{i,t}$, are invariant with respect to displacement in time [note that, by definition, mean values of random variables $\eta_{i,t}$ are equal to zero, $E(\eta_{i,t}) = 0$], i.e., $\text{Var}(\eta_{i,t}) = \text{Var}(\eta_i) = \sigma^2 > 0$, and $\text{Cov}(\eta_{i,t}, \eta_{i,t+s}) \neq 0$, for $s = 0, 1, ..., i = 1, 2, ..., N$, and integer valued t ($-\infty < t < +\infty$).

Since immediate computations are not possible and the firm's output-price decisions have to be made prior to the knowledge of the market price, the result computed in period t can be used only in subsequent periods, i.e., deviations $\eta_t + \Sigma_{i=1}^N \eta_{i,t}$, can be estimated based on the results computed in the past, and, consequently, always with certain error. It has to be stressed, however, that the variance of the error in the estimation increases with the time elapsed from observations of individual demands to the moment when decisions are made (see Radner and Van Zandt 1992, for a discussion). Therefore, the supplier faces not only the rather standard problem of finding appropriate estimations of demand but

---

[1]   In general, specifications of stochastic processes describing individual demands should also include a 'common noise' which could reflect aggregate demand shocks (i.e., which could equally affect all sources of demand), but to simplify the exposition we will disregard this common component.

[2]   A similar structure of demand was assumed by Radner and Van Zandt (1992).

also the problem of finding the optimal cost of these estimations, since data processing is inherently costly and the acquisition and analysis of more pieces of information (and in particular, more recent information) have to be weighed against the increasing costs of such an endeavour.

In general, the firm may find it advantageous to compute in subsequent periods (say, $t - m$, $m = 1, 2, ...$) deviations from the mean values of random variables $\eta_{i,t-m}$ coming from different subsets of sources (say, S_{t-m}, $m = 1, 2, ...$) and use them for the estimation of the total deviation from the expected demand in period t (rational strategy requires that sources of demand should be analysed cyclically one after the other).[3]

Denote the results computed in subsequent periods as $\eta_t^{S_{t-m}}, ..., \eta_t^{S_{t-1}}$. If the subsets $\{S_{t-m}, ..., S_{t-1}\}$ contain $n_{t-m}, ..., n_{t-1}$ ($n_{t-m} > 0, ..., n_{t-1} > 0$) sources of individual demand, then there exits integer number K ($N \geq K \geq 1$) such that $\sum_{i=1}^{K} n_{t-i} \leq N < \sum_{i=1}^{K+1} n_{t-j}$. Thus, the estimation of total deviation $\tilde{\eta}_t$ can be computed as

$$\tilde{\eta}_t = \sum_{i=1}^{K} \tilde{\eta}_t^{S_{t-i}} + \frac{N - \sum_{i=1}^{K} n_{t-i}}{n_{t-(K+1)}} \tilde{\eta}_t^{S_{t-(K+1)}} \qquad (2)$$

where $\tilde{\eta}_t^{S_{t-m}}$ is a forecast (for period t) of the sum of the deviations from the mean values of random variables coming from the sources included in the set S_{t-m} ($m = 1, 2, ..., K + 1$).

Since all the available predictions of partial deviations ($\tilde{\eta}_t^{S_{t-m}}$, $m = 1, 2, .., K + 1$) can be represented as linear combinations of the true values of corresponding partial deviations in past, the expected error in the prediction of total deviation equals zero. Furthermore, its variance (assuming that deviations from the expected values of individual demands, $\eta_{i,t}$, are independent, identically distributed, and time invariant) is

$$\sigma_t^2 = n_{t-i} \, \sigma_{t,i}^2 + \left(N - \sum_{i=1}^{K} n_{t-i} \right) \sigma_{t,K+1}^2 \qquad (3)$$

where

$$\sigma_{t,m}^2 = E\{[\eta_{i,t} - \tilde{\eta}_{i,t}(m)]^2\} \qquad (4)$$

is the variance of error in the estimation (with lag m, $m = 1, 2, ..., K + 1$) of the deviation of the random variable $\eta_{i,t}$ ($i = 1, 2, ..., N$) from its mean value, and $\tilde{\eta}_{i,t}(m)$ denotes the estimation with lag m ($m = 1, 2, ..., K + 1$) of the deviation of the random variable $\eta_{i,t}$ ($i = 1, 2, ..., N$) from its mean value.

---

[3] See Cukrowski (1996) for details.

The forecast of the inverse demand can be specified as $\tilde{P}_t(Q_t) = P(Q_t) + \tilde{\eta}_t$, where $P_t(Q_t) = a - bQ_t$ denotes the expected demand curve in period t ($-\infty < t < +\infty$). The prediction error $\tilde{\eta}_t$ is given by expression (2) and its variance σ_t^2 by an expression (3).

4   A Monopolistic Supplier and Retail Firms

Taking into account that the variability of demand decreases the quality of output-price decisions (i.e., output-price decisions deviate from the optimal decision that would be made if the variance were equal to zero) and that the results of demand analysis can be used only after the end of the period in which they were computed, the smallest variance of the prediction error corresponds to the case when all sources of demand are analysed in the preceding period. The analysis of the total demand in each period, however, requires a number of economic resources to be devoted to data processing in the firm, i.e., it induces significant costs that can not always be offset by the expected benefit from the output-price decision with a lower risk of error. Thus, instead of examining the demand coming from all sources in each period, the firm can sequentially analyse the demand coming from certain subsets of sources. In this case, however, the firm has to determine the optimal number of sources of demand that should be analysed in subsequent periods.

Suppose now that there are two types of firms operating in the market: a monopolistic supplier, s, of a single type of services and perfectly competitive retail firms, r, that can resell services and freely enter or exit the market. Suppose that firms (the supplier and retailers) are managed according to the wishes of their owners who are typical asset holders, and the decisions in each firm are made by a group of decision makers with sufficiently similar preferences to guarantee the existence of a group preference function, representable by a von Neuman-Morgen-stern utility function. Given these conditions we assume risk aversion, so that the utility functions of the supplier, U_s, and retail firms, U_r, are concave and differentiable functions of profits. The objective of both the supplier and the retail firms is to maximise the expected utility from profit (we assume that the firms set the volume of output supplied).

Assume that the firms are able to analyse market data and predict demand. Taking into account that the life of firms is unlimited, the optimisation task of firm x ($x \in \{s, r\}$) can be represented as the following infinite-horizon, discounted, dynamic programming problem:

$$\max_{Q_t, n_t} \sum_{t=0}^{\infty} \beta^t \, E\{U[\Pi_t(Q_t, \, \sigma_t, \, \eta^0, \, n_t)]\}, \tag{5}$$

where $\sigma_{x,t+1} = f(\sigma_{x,t}, n_{x,t})$, with $\sigma_{x,0} = \sigma_0 = (N\tilde{\omega}^2)^{1/2}$, E is an expectation operator, $U_x(\cdot)$, denotes the utility function of the firm $x \in \{s, r\}$, $\Pi_{x,}(\cdot)$, is the profit of the firm in the period t, $t = 0, 1, ..., (x \in \{s, r\})$. $Q_{x,t}$ is a quantity supplied by the firm $x(x \in \{s, r\})$ in the period t, $t = 0, 1, ...$; $\eta_{x,t}$ denotes the number of individual sources of demand analysed in firm x $(x \in \{s, r\})$ in the period t, $t = 0, 1, ...$; $\sigma_{x,t}$ is the standard deviation of the error in the prediction of the total deviation of the random variables $\eta_{i,t}$ $(i = 1, 2, ..., N)$ in firm x $(x \in \{s, r\})$ in the period t, $t = 0, 1, ...$; N is the total number of sources of demand, $\tilde{\omega}^2$ is the variance of the stochastic process underlying each individual demand around its mean, β is the discount factor, and $\beta \in (0, 1)$.

The cost of gathering and processing information in a given period and the benefits from this activity in future periods (i.e., smaller variance of the prediction error) specify the link which connects the present with the future. In other words, in the model considered, there is an intertemporal trade off between higher costs of data processing today and future benefits in the form of a higher expected utility. Thus, along the optimal path the disutility from the analysis of one additional source of demand in a period j $(j = 0, 1, ...)$ has to be equalised with the sum of the discounted marginal benefits in all future periods, i.e.,

$$-\frac{\partial E\{U_x[\Pi_{x,j}(Q_{x,j}, \sigma_{x,j}, n_{x,j})]\}}{\partial n_{x,j}} = \sum_{t=j+1}^{\infty} \beta^t \frac{\partial E\{U_x[\Pi_{x,t}(Q_{x,t}, \sigma_{x,t}, n_{x,t})]\}}{\partial \sigma_{x,t}} \frac{\partial \sigma_{x,t}}{\partial n_{x,t}}. \quad (6)$$

Assuming that all the parameters of the model are stationary over time, the optimal solution to an infinite-horizon, discounted, dynamic programming problem is time-invariant (see, e.g. Sargent 1987). Thus, in the problem considered, the optimal output and demand-predicting strategy is stationary, i.e., $Q_{x,0}^* = Q_{x,1}^* = Q_{x,2}^* = ... = Q_x^*$ and $n_{x,0}^* = n_{x,1}^* = n_{x,2}^* = ... = n_x^* (x \in \{s, r\})$. This implies that the optimal value of the standard deviation (σ_x^*) of the error in the prediction of the total deviation of the random variables $\eta_{i,t}$ $(i = 1, 2, ..., N)$ from their means is stationary and depends only on the number of individual demands analysed in every period, $\sigma_x^* = \sigma_x(n_x^*)$. It follows that the unique one-period cost of data processing can be related to each value of the standard deviation $\sigma_x(n_x^*)$, i.e., the costs of data processing in each period can be represented as a function of the standard deviation in the steady state, $g[\sigma_x(n_x^*)] \equiv V(n_x^*)$. Since the stationary standard deviation, $\sigma_x(n_x^*)$, decreases if the number of individual demands analysed in each period increases, the cost of data processing is a decreasing function of the steady state standard deviation. Note that differentiating the cost of data processing $V(n_x^*)$ with respect to n_x^* gives $dV(n_x^*)/dn_x^* = \partial g/\partial \sigma_x$ $d\sigma_x/dn_x^* > 0$. Since $dV(n_x^*)/dn_x^* > 0$ and $d\sigma_x/dn_x^* < 0$, it follows that $\partial g/\partial \sigma_x < 0$. Moreover, the shape of the function $V(n_x^*)$ $dV(n_x^*)/dn_x^* > 0$, $d^2V(n_x^*)/dn_x^{*2} > 0$ (see Cukrowski 1996), implies that g is a convex function of σ, i.e., $\partial^2 g[\sigma_x(n_x^*)]/\partial \sigma_x^2 > 0$.

Assume, for simplicity, that a standard deviation is the following function of the cost of data processing $\sigma_x = \sigma_0 e^{-\lambda g}$, where g denotes the cost of data processing, and λ $(\lambda > 0)$ is a parameter describing the current state of information processing technology. Consequently, for any $\sigma_x < \sigma_0$, the cost of data processing is specified as $g(\sigma_x) = -(\ln\sigma_x - \ln\sigma_0)/\lambda$.

The consideration above shows that the optimisation problem of the firm x ($x \in \{s, r\}$) can be solved in two steps. First, the optimal quantity Q_x^* and the optimal value of standard deviation σ_x^* can be determined, and, second, knowing σ_x^*, the optimal size of the cohorts of data summarised in each period can be found.

Thus, in the first stage the firm x ($x \in \{s, r\}$) chooses the steady-state quantity of output Q_x and the value of the standard deviation σ_x which maximise the following objective function

$$\max_{Q_x, \sigma_x}\{E\{U_x[\Pi_x(Q_x, \sigma_x)]\}. \tag{7}$$

To simplify the analysis, assume that the steady-state error in prediction of the total demand is a normally distributed random variable with zero mean and variance σ_x^2 (this corresponds to the case when random deviations follow stochastic processes with normally distributed random terms such as, for example, the autoregressive process of any order)[4]. Since the distribution of the total random deviation from the mean value of demand is normal, the total deviation can take positive and negative values, each having probability 0.5 (the expected value of positive deviation equals $\sigma_x/(2\pi)^{1/2}$ and the expected value of negative deviation equals $-\sigma_x/(2\pi)^{1/2}$.[5] Consequently, the total inverse random demand in any period t ($-\infty < t < +\infty$) can be approximated as $\tilde{P}(Q_x, \sigma_x) = a - bQ_x + \vartheta(\sigma_x)$ where $\vartheta(\sigma_x)$ is a random factor (not known ex ante) which with probability 0.5 equals $\gamma(\sigma_x)$ and with probability 0.5 equals $-\gamma(\sigma_x)$, where $\gamma(\sigma_x) = \sigma_x/(2\pi)^{1/2}$. Consequently, one can say that with probability 0.5 an inverse market demand is $\underline{P}(Q_x, \sigma_x) = a - bQ_x - \gamma(\sigma_x)$, and with probability 0.5 is $\overline{P}(Q_x, \sigma_x) = a - bQ_x + \gamma(\sigma_x)$. The expected market demand curve is determined as $P(Q_x) = a - bQ_x$.

Using this approximation, the optimisation problem of firm x ($x \in \{s, r\}$) can be represented as

$$\max_{Q_x, \sigma_x} [\Psi_x(Q_x, \sigma_x)] \equiv \max_{Q_x, \sigma_x} \{\tfrac{1}{2}U_x[\overline{\Pi}_x(Q_x, \sigma_x)] + \tfrac{1}{2}U_x[\underline{\Pi}_x(Q_x, \sigma_x)]\} \tag{8}$$

---

[4]   It should be stressed that, although the assumption of the normal distribution of the random deviations from the expected demand corresponds to the wide class of stochastic processes that would govern stochastic demand, it is chosen solely for simplicity and clarity, and no attempt is made at generality. We believe, however, that many of the qualitative results would hold also in the more general, and, consequently, more complicated models.

[5]   Expected values of positive and negative deviations are computed as $\int_0^\infty \tilde{\eta}(2\pi\sigma_x^2)^{-1/2} \exp(-\tilde{\eta}^2/2\sigma_x^2)\, d\tilde{\eta}$ and $\int_{-\infty}^0 \tilde{\eta}(2\pi\sigma_x^2)^{-1/2} \exp(-\tilde{\eta}^2/2\sigma_x^2)\, d\tilde{\eta}$, respectively.

where $\overline{\Pi}_x(Q_x,\sigma_x) \equiv Q_x\overline{P}(Q_x,\sigma_x) - F_x(Q_x,\sigma_x)$ and $\underline{\Pi}_x(Q_x,\sigma_x) \equiv Q_x\underline{P}(Q_x,\sigma_x)$ $- F_x(Q_x,\sigma_x)$, $F_x(Q_x,\sigma_x)$ denotes a cost function of the firm x.

5   Distribution of Output

First consider the case when a monopolistic supplier trades directly with final consumers. Assume that the supplier's cost function is $F_s(Q_s,\sigma_s) = c\,Q_s + g(\sigma_s) + B$, where Q_s denotes the volume of output produced, $g(\sigma_s)$ denotes the cost of data processing, c is the marginal cost, and B is the fixed cost. To simplify the analysis assume that the exact shape of the utility function U_s is specified as follows:

$$U_s(\Pi_s) = \begin{cases} u_1\,\Pi_s & \text{if}\quad \Pi_s < \Pi_s^0, \\ u_2\,\Pi_s + (u_1 - u_2)\Pi_s^0 & \text{if}\quad \Pi_s{}^3 \geq \Pi_s^0, \end{cases} \tag{9}$$

where $u_1 > u_2 > 0$ and $\underline{\Pi}_s < \Pi_s^0 < \overline{\Pi}_s$. [6]

The interior solution to the supplier's optimisation problem (see expression (8)) exists if

$$\lambda \geq \frac{8\,b}{(a-c)^2} \tag{10}$$

(see Appendix A for details).

Assuming that the primitives of the model: a, b, c, λ satisfy the condition above, the optimal steady state values of the volume of output supplied Q_s^* and the standard deviation of the demand σ_s^* are determined as

$$Q_s^* = \frac{a-c}{4\,b} + \sqrt{\left(\frac{a-c}{4\,b}\right)^2 - \frac{1}{2\,\lambda\,b}}, \tag{11}$$

$$\sigma_s^* = \left(\frac{a-c}{2} - \sqrt{\left(\frac{a-c}{2}\right)^2 - \frac{2\,b}{\lambda}}\right)\frac{\sqrt{2\,\pi}}{k_s}, \tag{12}$$

where $k_s = (u_1 - u_2)/(u_1 + u_2)$, $k_s \in (0, 1)$ for the risk-averse firm[7] (see Appendix A for the proof).

---

[6]   Note that a function defined is concave and twice differentiable if $\Pi_s \in (-\infty,\infty)/\Pi_s^0$.

[7]   Coefficient k_s characterises the attitude towards risk and increases with risk aversion.

Assume now that the monopolist can sell the output not to final consumers but to perfectly competitive retail firms which can freely enter and exit the market. Each individual retail firm operates in the market only if its expected utility from profit is at least equal to the utility of some benchmark activity φ ($\varphi \geq 0$). Since for a risk-averse retail firm earning random profit (Π_r) the following is true $U_r[E(\Pi_r)] > E[U_r(\Pi_r)] \geq \varphi \geq 0$, where U_r denotes the utility function of the retail firm and E is an expectation operator, the expected value of profit of a single retailer operating in the market is positive. This implies that the retail market can be established only if the expected value of the profit of the retail sector as a whole is positive, i.e., if the supplier sells services to retail firms at a lower price than the expected price to final consumers.

Proposition 1: *Rational behaviour of the risk-averse monopolistic supplier under uncertainty of demand implies that the supplier is always willing to sell services to retail firms at a lower price than the expected price to consumers.*

Proof. Under demand uncertainty the risk-averse monopolistic supplier trading directly with consumers earns random profit with the expected value $E[\Pi_s(Q_s^*, \sigma_s^*)]$, such that $E[\Pi_s(Q_s^*, \sigma_s^*)] < E[\Pi_s(Q_0^*, \sigma = 0)]$, where Q_0^* is the optimal monopolistic output without uncertainty (i.e., if $\sigma = 0$). Taking into account that the optimal volume of output supplied to consumers by retail firms, and, consequently, demanded from the supplier, is deterministic (see Appendix A), and that a risk-averse firm always prefers deterministic profit to random profit with the same (or even slightly higher) expected value, the deterministic price $P_0(Q_R)$ at which the supplier would be willing to sell the volume of output Q_R to retail firms should satisfy the following condition:

$$E\{U_s[P_0(Q_R)Q_R - c\,Q_R - B]\} \geq E\{U_s[\Pi(Q_s^*, \sigma_s^*)]\}. \tag{13}$$

Since, $E\{U_s[P_0(Q_R)Q_R - c\,Q_R - B]\} = U_s\{E[P_0(Q_R)Q_R - c\,Q_R - B]\}$ and for a risk-averse firm $U_s\{E[\Pi(Q_s^*, \sigma_s^*)]\} \geq E\{U_s[\Pi(Q_s^*, \sigma_s^*)]\}$, inequality (13) is satisfied for any $P_0(Q_R)$, such that[8]

$$P_0(Q_R)Q_R - c\,Q_R - B \geq E[\Pi(Q_s^*, \sigma_s^*)]. \tag{14}$$

Expression (14) states that the deterministic profit of the supplier (when it trades with retail firms) should be at least equal to the expected value of profit that the supplier would earn if he sold services directly to final demanders. Note that for any σ_s^*, $E[\Pi_s(Q, \sigma_s^*)]$ is a continuous, strictly concave, function of Q, positive for $Q \in (0, Q_C)$, where Q_C is the optimal competitive output without uncertainty, achieving its maximum for $Q = Q_0^*$. Since $E[\Pi_s(Q_s^*, \sigma_s^*)] < E[\Pi_s(Q_0^*, \sigma_s^*)]$, there exists an interval (say, (Q_A, Q_B), where $Q_A = Q_s^*$ and $Q_0^* < Q_B < Q_C$), in which

---

[8] Note that Q_R is a deterministic variable, and, consequently, is deterministic as well.

$E[\Pi_s(Q_s^*, \sigma_s^*)] < E[\Pi_s(Q, \sigma_s^*)]$. Plugging $E[\Pi_s(Q_s^*, \sigma_s^*)] = P_0(Q_R)\, Q_R - c\, Q_R - B$ and $E[\Pi_s(Q, \sigma_s^*)] = P(Q)\, Q - c\, Q - B - g(\sigma_s^*)$, where $P(Q)$ is an expected price if quantity Q is supplied to consumers, into the above inequality and rearranging, we get that $P_0(Q_R)\, Q_R < P(Q)\, Q - g(\sigma_s^*)$, and consequently, that $P_0(Q_R) < P(Q)$ for any $Q_R = Q \in (Q_A, Q_B)$. QED.

Suppose now that the supplier trades with retail firms, but it can not (or it is not legally allowed to) impose any vertical restraints, i.e., assume that the supplier is willing to sell any given volume of output Q_R at price $P_0(Q_R)$ to perfectly competitive retail firms. Assume also that there exists an interval, say (Q_α, Q_β), where $Q_A < Q_\alpha < Q_\beta < Q_B$, such that, for any volume of output supplied to retail market Q_r in this interval, retail market is organised (i.e., the number of retail firms operating in the market n is greater or equal to 1), and for any $Q \notin (Q_\alpha, Q_\beta)$ the retail market can not be organised ($n < 1$).

If the supplier trades with retail firms the cost function of a single retail firm is $F_r(Q_r) = Q_r\, P_0(Q_R) + g(\sigma_r)$, where Q_r is the volume of output supplied to final demanders by a single retail firm, $P_0(Q_R)$ is the price to retail firms if the volume of output Q_R is supplied to the retail market (to focus directly on the problem no additional cost is assumed), and $g(\sigma_r)$ is the steady state cost of data processing, which corresponds to the standard deviation σ_r. Consequently, the optimisation problem of each individual retail firm can be represented as[9]

$$\max_{Q_r, \sigma_r} E\{U_r[\Pi_r(Q_r, \sigma_r)]\} \equiv \max_{Q_r, \sigma_r}\{\tfrac{1}{2}U_r[\overline{\Pi}_r(Q_r, \sigma_r)] + \tfrac{1}{2}U_r[\underline{\Pi}_r(Q_r, \sigma_r)]\} \qquad (15)$$

where $\overline{\Pi}_r(Q_r, \sigma_r) \equiv Q_r[\overline{P}(n\, Q_r, \sigma_r) - P_0(n\, Q_r)] - g(\sigma_r)$, $\overline{P}(Q_r, \sigma_r) = a - b\, n\, Q_r + \gamma(\sigma_r)$, $\underline{\Pi}_r(Q_r, \sigma_r) \equiv Q_r[\underline{P}(n\, Q_r, \sigma_r) - P_0(n\, Q_r)] - g(\sigma_r)$, $\underline{P}(Q_r, \sigma_r) = a - b\, n\, Q_r - \gamma(\sigma_r)$, $\gamma(\sigma_r) = \sigma_r / (2\pi)^{1/2}$, σ_r denotes steady state standard deviation of total demand and Q_r denotes the output supplied to final demanders by a single retail firm (n is the number of retail firms operating in the market)[10].

To simplify the analysis assume that the exact shape of the utility function U_r is specified as follows:

$$U_r(\Pi_r) = \begin{cases} y_1\, \Pi_r & \text{if} \quad \Pi_r < \Pi_r^0, \\ y_2\, \Pi_r + (y_1 - y_2)\Pi_r^0 & \text{if} \quad \Pi_r \geq \Pi_r^0, \end{cases} \qquad (16)$$

where $y_1 > y_2 > 0$ and $\underline{\Pi}_r < \Pi_r^0 < \overline{\Pi}_r$ [11].

---

[9] Note that if the optimal value supplied by each individual retail firm (Q_r^*) exists, it is not a random but a deterministic variable.

[10] Recall that under uncertainty of demand the number of firms operating in perfectly competitive market is finite (see Ghosal 1996, for empirical evidence).

[11] Note that a function defined is concave and twice differentiable if $\Pi_r \in (-\infty, \infty) / \Pi_r^0$.

Thus each individual retail firm considers the maximisation problem (8). The interior solution to this optimisation problem exists if

$$\lambda \geq \frac{8\,b\,n}{(a-c)^2} \tag{17}$$

(see Appendix). Assuming that the primitives of the model satisfy inequality (17), the optimal steady state values of the volume of output supplied to final consumers by each individual retail firm Q_r^* and steady state value of the standard deviation of demand σ_r^* are determined as

$$Q_r^* = \frac{a-c}{4\,b\,n} + \sqrt{\left(\frac{a-c}{4\,b\,n}\right)^2 - \frac{1}{2\,\lambda\,b\,n}} \tag{18}$$

$$\sigma_r^* = \left(\frac{a-c}{2} - \sqrt{\left(\frac{a-c}{2}\right)^2 - \frac{2\,b\,n}{\lambda}}\right)\frac{\sqrt{2\,\pi}}{k_r} \tag{19}$$

where $k_r = (y_1 - y_2)/(y_1 + y_2)$, $k_r \in (0, 1)$ for risk-averse firms[12] (see Appendix for the proof).

Since the expected value of profit of each retail firm operating in the market is positive and the maximum expected value of profit of the retail sector equals $E[\Pi_s(Q_0^*, \sigma = 0)] - E[\Pi_s(Q_s^*, \sigma_s^*)]$ where Q_0^* is the optimal monopolistic output without uncertainty (i.e., if $\sigma_0 = 0$), in equilibrium only a finite number of perfectly competitive retail firms can operate in the market. Assuming that the number of firms n is a continuous number instead of an integer, in market equilibrium the expected utility of profit of a single retail firm is

$$E\{U_r[\Pi_r(Q_r^*, \sigma_r^*)]\} \geq \varphi \geq 0. \tag{20}$$

Since, for the risk-averse firm $U_r\{E[\Pi_r(Q_r^*, \sigma_r^*)]\} \geq E\{U_r[\Pi_r(Q_r^*, \sigma_r^*)]\}$ the inequality above is satisfied only if $U_r\{E[\Pi_r(Q_r^*, \sigma_r^*)]\} \geq 0$, i.e., if $E[\Pi_r(Q_r^*, \sigma_r^*)] > 0$. Taking into account that $E[P(n\,Q_r^*, \sigma_r^*)] - P_0(n\,Q_r^*) > 0$ if $n\,Q_r^* \in (Q_A, Q_B)$, where $Q_A = Q_s^*$ and $Q_0^* < Q_B < Q_C$ (Q_C is a perfectly competitive output), the condition (20) is satisfied only if $Q_s^* < n\,Q_r^* < Q_C$, i.e., if the quantity of output supplied through the retail market is greater than it would be if the supplier traded directly with final consumers.

An important implication of the result above is that retail trade under uncertainty of demand changes the distribution of welfare in the economy. In particular, it decreases the expected value of the deadweight loss (i.e., the volume

---

[12] Coefficient k_r characterises the attitude towards risk and increases with risk aversion.

of output is higher than it would be without the retail market) and increases the expected value of consumer surplus (consumers consume more and at a lower price). The monopolistic supplier is also better off since the supplier changes random profit to deterministic profit with the same expected value.

6 Progress in Information Processing Technology and the Size of Retail Markets

Since the optimal volumes of output supplied to final consumers by both the monopolistic supplier and the retail firms (see expressions (11) and (18), respectively), as well as the steady state values of the standard deviations of the demand (expressions (12) and (19)) depend on the parameter describing the current state of information processing technology λ, technological progress in data processing, which makes predictions cheaper (increases λ), changes the characteristics of market equilibrium. The optimal output of the monopolistic supplier shifts closer to the optimal monopolistic output without uncertainty of demand, and consequently, the offer curve to retail firms $P_0(Q_R)$ shifts upward. This decreases the difference between the expected price to consumers and the price to retail firms. The expected profit of the retail sector as a whole and the expected profit of each particular retail firm both decrease, and, consequently, the number of retail firms operating in the market tends to decrease. At the same time retail firms are also able to make better predictions (i.e., decrease the variance of demand), and, as a result, are able to increase the expected value of profit for any particular volume of output supplied. Therefore, other things being constant, more retail firms can operate in the market. The total effect of technological change on the number of firms in market equilibrium is characterised by the proposition below.

Proposition 2: *Technological progress in information processing increases the number of retail firms operating in the market.*

Proof. Note that for any fixed λ both Q_r^* and s_r^* can be considered as functions of n. Assume for the time being that the number of retail firms n is continuous (rather than integer valued) and consider function

$$G(\sigma_r^*(n), Q_r^*(n)) \equiv -\varphi^* + \Psi_r(\sigma_r^*(n), Q_r^*(n)), \tag{21}$$

where $\varphi^* \geq \varphi \geq 0$. Taking into account that in market equilibrium the expected utility from profit of each individual retailer operating in the market must be at least equal to φ ($\varphi \geq 0$), in market equilibrium $G(\sigma_r^*(n), Q_r^*(n)) \equiv 0$. Rearranging the latest expression we get

$$Q_r^*(n)[a - b\, n\, Q_r^*(n) - c] \;-\; \frac{E[\Pi_s(Q_s^*)] + B}{n} \;-\; \frac{\ln\sigma_0 - \ln\sigma_r^*(n)}{\lambda} \;\equiv$$

$$\frac{2\varphi^*}{y_1 + y_2} + \frac{(k_r\,\sigma_r^* n)}{\sqrt{2\pi}} - 2k_r\Pi_r^0. \tag{22}$$

The left hand side of the expression (22) can be represented as

$$\left(Q_r^*(n)[a - b\, n\, Q_r^*(n) - c] \;-\; \frac{\ln\sigma_0 - \ln\sigma_r^*(n)}{\lambda} \right) - \frac{B}{n} - \frac{1}{n}\, E[\Pi(Q_s^*)]. \tag{23}$$

The numerator in the expression above describes total expected profit of the retail sector. Therefore, expression (23) describes expected profit of a single retail firm, which in equilibrium has to be equal to $\chi > 0$. Consequently, in market equilibrium

$$\chi^* \;\equiv\; \frac{2\,\varphi^*}{y_1 + y_2} + \frac{k_r\,\sigma_r^*(n)}{\sqrt{2\pi}} - 2\,k_r\,\Pi_r^0. \tag{24}$$

Taking into account expression (19) and rearranging an equilibrium condition can be represented as

$$\left(\frac{2\,\varphi^*}{y_1 + y_2} - \chi \right) + \frac{a - c}{2} - \sqrt{\left(\frac{a - c}{2} \right)^2 - \frac{2\, b\, n}{\lambda}} - 2\, k_r\, \Pi_r^0 \;\equiv\; 0. \tag{25}$$

Assume now that λ is not constant but can fluctuate. Define a function

$$H(\lambda, n) \;\equiv\; \left(\frac{2\,\varphi^*}{y_1 + y_2} - \chi \right) + \frac{a - c}{2} - \sqrt{\left(\frac{a - c}{2} \right)^2 - \frac{2\, b\, n}{\lambda}} - 2\, k_r\, \Pi_r^0 \;\equiv\; 0. \tag{26}$$

Since $H(\lambda, n)$ is a continuously differentiable function of λ and n, according to the implicit function theorem, the first derivative of n with respect to λ is $dn/d\lambda = -(\partial H / \partial \lambda)/(\partial H / \partial n)$. $\partial H / \partial n$ is always positive, and $\partial H / \partial \lambda$ is negative. Therefore, $\partial n / \partial \lambda > 0$, i.e., the number of retail firms operating in the market increases with λ (i.e., with technological progress in data processing). QED.

7   Summary and Conclusion

The purpose of this contribution was to analyse the potential impact of technological progress in information processing as evidenced, for example, by data mining and computational intelligence technologies such as neurocomputing on the size of retail markets. The analysis focused on a single commodity market with uncertain demand. It has been shown that demand uncertainty can be considered an independent source of retail trade, and, consequently, the ability of firms to process information and predict demand may affect the characteristics of retail markets. The results derived show that risk-averse firms always devote resources to demand forecasting, producers are better off trading with retailers than with final consumers, and the volume of output supplied through the retail market is always greater than it would be if producers traded directly with consumers (i.e., it increases welfare). Finally, we proved that technological progress in information processing (that improves predictions and/or makes them cheaper, decreases uncertainty about demand in retail firms much more than in the supplier's firm) increases the size of retail markets.

An important implication of the results derived in the paper is that technological progress in information processing increases the number of retail firms operating in the market and, thus, opens up new job opportunities for workers who will loose their jobs in old resource intensive industries in the transition process from an industrial to an information-based economy.

Acknowledgement: The first author gratefully acknowledges the grant provided by the Institute for Urban and Regional Research at the Austrian Academy of Sciences.

References

Cukrowski J. (1996): Demand uncertainty, forecasting and monopolistic equilibrium, CERGE-EI Working Paper 97, Center for Economic Research, Prague

De Canio S.J. and Watkins W.E. (1998): Information processing and organisational structure, *Journal of Economic Behavior and Organisation* 36 (3), 275-294

Ghosal V. (1996): Does uncertainty influence the number of firms in an industry? *Economics Letters* 50, 229-223

Heal G. (1980): Spatial structure of the retail trade: A study in product differentiation with increasing returns, *The Bell Journal of Economics* 11 (2), 565-583

Jeannet J.P. and Hennessey H.D. (eds.) (1995): *Global Marketing Strategies*, Houghton Mifflin Company, Boston

Leland H.E. (1972): Theory of the firm facing uncertain demand, *American Economic Review* 62, 278-291

Lim C. (1980): Ranking behavioral modes of the firm facing uncertain demand, *American Economic Review* 70 (1), 217-224

Nelson R.R. (1961): Uncertainty, prediction, and competitive equilibrium, *Quarterly Journal of Economics* 75, 41-62

Radner R. and Van Zandt T. (1992): Information processing in firms and return to scale, *Annales d'Economie et de Statistique* 25/26, 265-298

Sandmo A. (1971): On the theory of the competitive firm under price uncertainty, *American Economic Review* 61, 65-73

Sargent T.J. (1987): *Dynamic Macroeconomic Theory*, Harvard University Press, Cambridge [MA]

Webster F. (1994): What information society, *The Information Society* 10, 1-23

Wilson R. (1975): Informational economies of scale, *The Bell Journal of Economics* 6 (1), 184-195

Appendix A: The Maximisation Problem of the Supplier

The objective function of the monopolistic supplier trading with final demanders can be approximated as

$$
\max_{Q_s, \sigma_s} \Psi_s(Q_s, \sigma_s) \equiv \tfrac{1}{2} u_1 \left[Q_s \left(a - b Q_s - \frac{\sigma_s}{\sqrt{2\pi}} \right) - c Q_s - B - \left(\frac{\ln \sigma_0}{\lambda} - \frac{\ln \sigma_s}{\lambda} \right) \right]
$$
$$
+ \tfrac{1}{2} \left\{ u_2 \left[Q_s \left(a - b Q_s + \frac{\sigma_s}{\sqrt{2\pi}} \right) - c Q_s - B - \left(\frac{\ln \sigma_0}{\lambda} - \frac{\ln \sigma_s}{\lambda} \right) + \left(u_1 - u_2 \right) \Pi_s^0 \right] \right\},
\tag{A.1}
$$

where Q_s denotes the volume of output supplied, and σ_s is the steady state standard deviation of demand. The first order conditions to the above optimisation problem can be represented as

$$
\frac{\partial \Psi_s(Q_s, \sigma_s)}{\partial Q_s} = \tfrac{1}{2} u_1 \left(a - 2 b Q_s - \frac{\sigma_s}{\sqrt{2\pi}} - c \right) + \tfrac{1}{2} u_2 \left(a - 2 b Q_s + \frac{\sigma_s}{\sqrt{2\pi}} - c \right) = 0 \tag{A.2}
$$

$$
\frac{\partial \Psi_s(Q_s, \sigma_s)}{\partial \sigma_s} = \tfrac{1}{2} u_1 \left(- \frac{Q_s}{\sqrt{2\pi}} + \frac{1}{\lambda \sigma_s} \right) + \tfrac{1}{2} u_2 \left(\frac{Q_s}{\sqrt{2\pi}} + \frac{1}{\lambda \sigma_s} \right) = 0 . \tag{A.3}
$$

The second order conditions to this maximisation problem require the Hessian of the objective function to be negative definite (it guarantees that the objective function is strictly concave). This Hessian is negative-definite (the objective function is strictly concave) iff

$$
\frac{\partial^2 \Psi_s(Q_s, \sigma_s)}{\partial Q_s^2} < 0 \tag{A.4}
$$

and

$$\frac{\partial^2 \Psi_s(Q_s, \sigma_s)}{\partial Q_s^2} \frac{\partial^2 \Psi_s(Q_s, \sigma_s)}{\partial \sigma_s^2} - \left(\frac{\partial^2 \Psi_s(Q_s, \sigma_s)}{\partial Q_s \partial \sigma_s}\right)^2 > 0. \tag{A.5}$$

Taking derivatives and rearranging, we conclude that the second order conditions are satisfied iff

$$\lambda < b \left[k_s \sigma_s (8\pi)^{-\frac{1}{2}}\right]^{-2}. \tag{A.6}$$

Rearranging the first order conditions, we obtain two possible values of output which maximise the objective function considered

$$Q_{s,1} = \frac{a-c}{4b} + \sqrt{\left(\frac{a-c}{4b}\right)^2 - \frac{1}{2\lambda b}} \tag{A.7}$$

$$Q_{s,2} = \frac{a-c}{4b} - \sqrt{\left(\frac{a-c}{4b}\right)^2 - \frac{1}{2\lambda b}}, \tag{A.8}$$

assuming that $[(a-c)/4b]^2 - (2\lambda b)^{-1} \geq 0$, i.e.,

$$\lambda \geq 8b(a-c)^{-2}. \tag{A.9}$$

If the cost of data processing goes to zero (λ goes to infinity), the firm knows demand almost perfectly, and the optimal volume of output goes to optimal volume of output without uncertainty $(a-c)/2b$. Consequently, the optimal quantity of output supplied Q_s^* is determined by the first expression, i.e., $Q_s^* = Q_{s,1}$.

Similarly, rearranging the first order conditions, we get two possible values of steady state standard deviation which maximise the objective function considered[13]

$$\sigma_{s,1} = \left(\frac{a-c}{2} - \sqrt{\left(\frac{a-c}{2}\right)^2 - \frac{2b}{\lambda}}\right) \frac{\sqrt{2\pi}}{k_s} \tag{A.10}$$

$$\sigma_{s,2} = \left(\frac{a-c}{2} + \sqrt{\left(\frac{a-c}{2}\right)^2 - \frac{2b}{\lambda}}\right) \frac{\sqrt{2\pi}}{k_s}. \tag{A.11}$$

---

[13] The square root in the expressions (A.10) and (A.11) is non-negative if $\lambda \geq 8b/(a-c)^2$.

Here again, if the cost of data processing goes to zero (λ goes to infinity), the firm knows demand almost perfectly (σ_s goes to zero). Consequently, the optimal value of the standard deviation σ_s^* is determined by the expression (A.10), i.e., $\sigma_s^* = \sigma_{s,1}$. Taking into account condition (A.9), plugging (A.10) into inequality (A.6) and rearranging we get the following condition

$$
2\,b\left(\frac{a-c}{2}\right)^{-2} \le \lambda < 4\,b\left(\frac{a-c}{2} - \left[\left(\frac{a-c}{2}\right)^2 - \frac{2\,b}{\lambda}\right]^{\frac{1}{2}}\right)^{-2}
\tag{A.12}
$$

which is always satisfied if $\lambda \ge 8b/(a-c)^2$ (i.e., Q_s^* and σ_s^* corresponds to the maximum of the objective function).

Appendix B: The Maximisation Problem of the Retail Firm

The objective function of the retail firm can be approximated as

$$
\max_{\sigma_r, Q_r} \Psi_r(\sigma_r, Q_r) \equiv \tfrac{1}{2} y_1 \left[Q_r\left(a - b\,n\,Q_r - \frac{\sigma_r}{\sqrt{2\,\pi}}\right) - P_0(n\,Q_r)Q_r - \left(\frac{\ln\sigma_0}{\lambda} - \frac{\ln\sigma_r}{\lambda}\right) \right]
$$

$$
+ \left\{ \tfrac{1}{2} y_2 \left[Q_r\left(a - b n Q_r + \frac{\sigma_r}{\sqrt{2\pi}}\right) - P_0(n Q_r)Q_r - \left(\frac{\ln\sigma_0}{\lambda} - \frac{\ln\sigma_r}{\lambda}\right) \right] + (y_1 - y_2)\Pi_r^0 \right\}
\tag{B.1}
$$

where Q_r denotes the volume of output supplied, $P_0(n\,Q_r) = E(Q_s^*) + B/(n\,Q_r) + c$, and σ_r is the steady state standard deviation of demand. The first order conditions to the above optimisation problem can be represented as

$$
\frac{\partial \Psi_r(Q_r, \sigma_r)}{\partial Q_r} = \tfrac{1}{2} y_1 \left(a - 2\,b\,n\,Q_r - \frac{\sigma_r}{\sqrt{2\,\pi}} - c \right) +
$$

$$
+ \tfrac{1}{2} y_2 \left(a - 2\,b\,n\,Q_r + \frac{\sigma_r}{\sqrt{2\,\pi}} - c \right) = 0,
\tag{B.2}
$$

$$
\frac{\partial \Psi_r(Q_r, \sigma_r)}{\partial \sigma_r} = \tfrac{1}{2} y_1 \left(-\frac{Q_r}{\sqrt{2\,\pi}} + \frac{1}{\lambda\,\sigma_r} \right) + \tfrac{1}{2} y_2 \left(\frac{Q_r}{\sqrt{2\,\pi}} + \frac{1}{\lambda\,\sigma_r} \right) = 0 .
\tag{B.3}
$$

The second order conditions to this maximisation problem require the Hessian of the objective function to be negative definite (it guarantees that the objective function is strictly concave). This Hessian is negative-definite (the objective function is strictly concave) iff

$$\frac{\partial^2 \Psi_r(Q_r, \sigma_r)}{\partial Q_r^2} < 0 \tag{B.4}$$

and

$$\frac{\partial^2 \Psi_r(Q_r, \sigma_r)}{\partial Q_r^2} \frac{\partial^2 \Psi_r(Q_r, \sigma_r)}{\partial \sigma_r^2} - \left(\frac{\partial^2 \Psi_r(Q_r, \sigma_r)}{\partial Q_r \partial \sigma_r} \right)^2 > 0. \tag{B.5}$$

Taking derivatives and rearranging we get that the second order conditions are satisfied iff

$$\lambda < n \, b \left(k_r \, \sigma_r / \sqrt{8\pi} \right)^{-2} \tag{B.6}$$

Rearranging the first order conditions we get two possible values of output which maximise the objective function considered

$$Q_{r,1} = \frac{a-c}{4 \, b \, n} + \sqrt{\left(\frac{a-c}{4 \, b \, n} \right)^2 - \frac{1}{2 \, \lambda \, b \, n}} \tag{B.7}$$

$$Q_{r,2} = \frac{a-c}{4 \, b \, n} - \sqrt{\left(\frac{a-c}{4 \, b \, n} \right)^2 - \frac{1}{2 \, \lambda \, b \, n}}, \tag{B.8}$$

assuming that $[(a-c)/4 \, b \, n]^2 - (2 \, \lambda \, b \, n)^{-1} \geq 0$, i.e.,

$$\lambda \geq 8 \, b \, n \, (a-c)^{-2}. \tag{B.9}$$

If cost of data processing goes to zero (λ goes to infinity) the firm knows demand almost perfectly, and the optimal volume of output goes to the optimal volume of output without uncertainty $(a-c)/2 \, b \, n$. Consequently, the optimal quantity of output supplied Q_r^* is determined by the first expression, i.e., $Q_r^* = Q_{r,1}$. Similarly, by rearranging the first order conditions, we get two possible values of steady state standard deviation which maximise the objective function considered[14]

---

[14] The square root in the expressions (B.10) and (B.11) is non negative if $\lambda \geq 8 \, b \, n$ $(a-c)^2$.

$$\sigma_{r,1} = \left(\frac{a-c}{2} - \left[\left(\frac{a-c}{2} \right)^2 - \frac{2 b n}{\lambda} \right]^{\frac{1}{2}} \right) \frac{(2\pi)^{\frac{1}{2}}}{k_r} , \tag{B.10}$$

$$\sigma_{r,2} = \left(\frac{a-c}{2} - \left[\left(\frac{a-c}{2} \right)^2 - \frac{2 b n}{\lambda} \right]^{\frac{1}{2}} \right) \frac{(2\pi)^{\frac{1}{2}}}{k_r} . \tag{B.11}$$

If cost of data processing goes to zero (λ goes to infinity) the firm knows demand almost perfectly (σ_r goes to zero). Consequently, the optimal value of the standard deviation σ_r^* is determined by the expression (B.10), i.e., $\sigma_r^* = \sigma_{r,1}$. Taking into account condition (B.9), plugging (B.10) into inequality (B.6) and rearranging we get the following condition

$$2 b n \left(\frac{a-c}{2} \right)^{-2} \leq \lambda < 4 b n \left(\frac{a-c}{2} - \left[\left(\frac{a-c}{2} \right)^2 - \frac{2 b n}{\lambda} \right]^{\frac{1}{2}} \right)^{-2} \tag{B.12}$$

which is always satisfied if $\lambda \geq 8 b n / (a-c)^2$ (i.e., Q_r^* and σ_r^* correspond to the maximum of the objective function).

Part II

Innovation and Network Activities

5 The New Economy and Networking

This chapter connects the knowledge based economy with the formation of networks, a field that constitutes a moving target. Networks have certainly not ceased to develop, if anything it is the hierarchical organisation that is in retreat. The literature reviewed in this chapter reflects this state of things. It is not a literature on phenomena that has reached equilibrium, but one that is vigorously developing. Clearly, to review such an expanding field constitutes an almost impossible task, at least for what concerns completeness of coverage. Thus, the author has deliberately chosen to address the main problems and to cover the main classes of literature. The main points about the economics of knowledge are stated in Section 2, and the literature on networks is surveyed in Sections 3 to 5. The final section points to the new role of government that is emerging as the knowledge-driven and globalising economy alters the way firms organise their operation within and across nation states.

1 Introduction

There is very little disagreement that we are entering into a new stage of market-based capitalism, a stage in which the economy is more strongly and more directly rooted in the production, distribution and use of knowledge than ever before. Knowledge creation and diffusion are the key driving forces of long-term economic growth, and the primary basis for competitiveness. Knowledge is an economic resource in its own right, taking the privileged role that once was accorded to natural resources. The dynamics of the economy are coming to rely more and more on learning or investments in knowledge creation and less on investments in physical capital. Knowledge creation through which change is anticipated and even initiated on a continuous base is becoming the vehicle to achieve sustained competitive advantage in the economy. These characteristics make it legitimate to speak of a new era – the *knowledge based economy*, the *knowledge-driven economy* or the *learning economy* – accompanied by a process of globalisation that increases pressures and expectations.

The knowledge based economy 'has emerged in the last quarter of the twentieth century', as Manuel Castells (1996, p. 66) claims, 'because the Information Technology Revolution provides the indispensable, material basis for such a new economy. It is the historical linkage between the knowledge-information base of the economy, its global reach, and the Information Technology Revolution that gives a new distinctive economic system'. No doubt, the *information technology revolution* and the knowledge based economy are closely interrelated. The convergence of computing, information and telecommunication technologies has drastically changed the conditions for the production and dissemination of knowledge as well as its coupling to the production system. New flexible information and communication technologies such as Internet, Web, intranet, extranet, data warehousing and data mining, as well as collaborative groupware technologies, are responsible for major changes in our abilities to handle data and information, to codify knowledge and to transmit codified knowledge. But economists have increasingly recognised that tacit rather than codified knowledge is vital to the innovation process, and tacit knowledge is embedded in people and not in information systems.

One of the most fundamental characteristics of the current phase of the knowledge based economy is the growing extent to which actors need to co-operate more actively and more purposefully with each other in order to cope with increasing market pressures stemming from globalisation, liberalisation of markets and changing patterns of demand. This is evidenced by both, the growth of interfirm networks and the closer integration of research, development, production and marketing within the company. Firms co-operate in order to gain rapid access to new technologies or markets, to benefit from economies of scale in joint R&D and production, to tap into external sources of know-how, and to share risks. The need for co-operation essentially stems from the specific nature of knowledge – as distinct from information – in particular its tacit and specific features.

The current chapter attempts to connect the knowledge based economy with the formation of networks, a field that constitutes a moving target. Networks have certainly not ceased to develop, if anything it is the hierarchical organisation that is in retreat. The literature reflects this state of things. It is not a literature on a phenomenon that has reached an equilibrium, but one that is still vigorously developing. Clearly, to review such an expanding field constitutes an almost impossible task, at least for what concerns completeness of coverage. Thus, we have deliberately chosen to address the main problems and to cover the main classes of literature. The main points about the economics of knowledge are stated in the section that follows, and the literature on networks is surveyed in Section 3 until Section 5. The final section points to the new role of government that is emerging as the knowledge-driven and globalising economy alters the way firms organise their operation within and across nation states.

2   Creation and Diffusion of Knowledge

In order to get a deeper understanding of the role of knowledge in the new economy in general and the processes of knowledge generation and diffusion in particular, it is useful to draw our attention to the following:

(a) the *complex nature of knowledge,*

(b) concepts such as the notions of *tacit* and *codified knowledge,* which have been available for a while,

(c) new concepts such as *knowledge conversion, absorption capacity, innovative capacity*, and *knowledge spillovers*, which lie at the root of the formation of formal or informal networks.

2.1 On the Nature of Knowledge

Recognition of the importance of linkages and relationships in everyday innovation processes has led to the development of a vital distinction between *knowledge* and *information*. The distinction between the two, however, is less clear-cut. However, there are several characteristics that differentiate knowledge and information. First, information characteristically refers to specific situations, conditions, processes, or objects. As such, it contains a level of detail and specificity that makes it appropriate to the given task. Beyond that, information alone becomes less valuable unless it is transformed into knowledge. Second, information contains data in a context and is generally limited to the context in which it was created. Knowledge transcends the specific context of the information and enables it to be applied to a variety of situations. Third, information is based on time and, thus, changes continually. Knowledge gleaned from yesterday's and today's information can be utilised to understand tomorrow's information.

In summary, information is factual and typically provides only for the *what* of the situation, whereas knowledge is characteristically complex and aims to discover the *why* (procedural knowledge) and *how* (skills and competences). Information may lead to knowledge when its value is somewhat enhanced through interpretation, organisation, filtration, selection, or engineering, utilising, for example, some more sophisticated statistical and modelling tools such as data mining. Knowledge is not only more complex, but establishes also – as Pier Paolo Saviotti (1998, p. 845) phrases it 'generalisations and correlations between variables. Knowledge is, therefore, a *correlational structure* ... Each piece of knowledge (e.g., a theory) establishes correlations over some variables and over particular ranges of their values' (emphasis added).

Furthermore, knowledge is a *retrieval-interpretive* structure insofar that particular pieces of information can be understood only in the context of a given type of knowledge and by those who possess that knowledge. Knowledge is also *cumulative* in nature. This feature stresses the very important learning processes in the development and use of new knowledge, both by firms initiating innovations and by those approaching the new technology at later points in the development via diffusion processes. These learning processes include learning-by-doing (for example, increasing the efficiency of production systems), learning-by-using (for example, increasing the efficiency by using complex systems), and learning-by-interacting involving users and producers in an interaction resulting in product innovations.

The *knowledge base* of a corporation contains knowledge in all its forms, for example, from simple and routine procedures of everyday production activities to shared practices and habits of thought, and from embodied to disembodied knowledge. Embodied knowledge is typically found in a product, with features that can be easily revealed by taking the product apart. Reverse engineering performed on a given product can reveal such features. Disembodied knowledge, on the other hand, tends to be a process and is usually abstract or complex in formulation, with certain degrees of tacitness. It involves new ways of doing things and may require a great deal of experience to be revealed.

2.2 Tacit and Codified Knowledge

Knowledge can be tacit or codified in the form of publications, patents etc. It is always at least partly tacit in the minds of those who create it. 'Tacitness refers to those elements of knowledge, insight and so on, that individuals have', as Dosi (1988, p. 1126) has suggested on the basis of earlier insights by Polanyi (1967), 'which are ill-defined, uncodified and unpublished, which they themselves can not fully express and which differ from person to person, but which may to some significant degree be *shared by collaborators and colleagues* who have a common experience' (emphasis added). Tacit knowledge can take many forms, such as skills and competences, specific to individuals or to groups of co-operating individuals, shared beliefs and modes of interpretation, but is not codified or possibly uncodifiable.

Much of the essential knowledge is specialised and resides in tacit form within experienced individual researchers or engineers. Among the most important of these non-codifiable skills are the acquisition and effective utilisation of knowledge. This person-embodied expertise is generally difficult to transfer and can often only be passed on effectively by personalised and generally localised apprenticeship-like relationships or by physical transfer of people who are carriers of the knowledge.

Codified scientific and technological knowledge is complementary to more tacit forms of knowledge within corporations. Part of this knowledge may become codified in time. The process of codification involves the gradual convergence of the scientific community and other users on common concepts and definitions and on common contents and theories. This process requires time and costs, but is becoming increasingly supported by novel capture and collection technologies.

2.3 Knowledge Conversion

Knowledge is produced by individuals and not by the organisation itself. Knowledge conversion addresses the problem of pulling knowledge into the organisation's knowledge base, where it can then be managed and shared. Otherwise, this knowledge remains highly personal and difficult to communicate.

Nonaka and Takeuchi (1995) have developed a simple, but elegant model to account for the generation of knowledge in the organisation. What they call the 'knowledge-creating company' is based on the organisational interaction between codified and tacit knowledge. The interaction between these two forms of knowledge is the key dynamics of knowledge creation in the business organisation and brings about the following four phases of knowledge conversion, each requiring special learning processes:

(a) *Externalisation*: The conversion from tacit knowledge to explicit (i.e., codified) knowledge, usually by articulating tacit knowledge and turning it into an explicit form, such as a report or documentation (codification is at the heart of this mode). Collaborative technologies such as groupware may be used to capture the tacit knowledge of individuals or teams after it has been made explicit in a document or through a forum or on-line discussion.

(b) *Socialisation*: The conversion from tacit knowledge to tacit knowledge through sharing experiences, imitation and practice. This type of conversion occurs through coaching, in apprenticeships, at seminars and conferences, or simply during employee interaction.

(c) *Internalisation*: The conversion from explicit knowledge back to tacit knowledge. This type of conversion is closely related to learning-by-doing and leads to operational-procedural knowledge.

(d) *Dissemination*: The conversion from explicit knowledge to explicit knowledge by the owners' sharing it with one another. Dissemination is the primary way knowledge is leveraged throughout the organisation and may rely on technologies such as computer-based training, groupware, and data warehouses.

By means of these phases of knowledge conversion, not only is worker experience communicated and amplified to increase the formal body of the knowledge base of the organisation, but also knowledge generated in the outside world can be incorporated into the tacit habits of workers to enable them to improve on the standard procedures. This process requires, as Castells (1996) emphasised, appropriate policies to motivate workers to share and not to keep knowledge solely for their own profit. It also requires stability of the labour force in the organisation because only then it does become rational for the organisation to diffuse explicit knowledge and for the individual to share his knowledge with the organisation.

2.4 The Spread of Knowledge

Firms need to absorb, create and exchange knowledge interdependently. In other words, knowledge creation and diffusion usually emerge as a result of an interactive and collective process within a web of personal and institutional connections that evolve over time. Knowledge transfer may occur through equipment-embodied or -disembodied diffusion. The former is the process by which knowledge spreads in the economy through the purchase of technology-intensive equipment. Disembodied knowledge diffusion refers to the process whereby knowledge diffuses through other channels. This type of knowledge diffusion may occur via reverse engineering (using computer-aided design and manufacturing applications) by a firm of its rivals' products, descriptions of new products, or production processes found in catalogues, publications, or patent applications. It can also be the byproduct of mergers and acquisitions, joint ventures, or other forms of interfirm co-operation. However, labour mobility among research institutions and the business sector as well as among firms is the most important channel for transmitting tacit knowledge. The extent to which knowledge flows through these different channels depends upon the capability of the recipient firm, the nature of the knowledge itself, and other factors.

Two notions are central to the understanding of disembodied knowledge diffusion: the first is that of absorption capacity and the second that of knowledge spillovers. The *absorption capacity* of firms and other organisations, as expressed by Cohen and Levinthal (1989), refers to the ability to learn, assimilate and use knowledge developed elsewhere through a process that involves substantial investments, especially of an intangible nature. This capacity depends crucially on the learning experience, which in turn may be enhanced by in-house R&D activities. The concept of absorption capacity implies that, in order to have access to a piece of knowledge developed elsewhere, it is necessary to have undertaken R&D on something similar. Thus, R&D may be viewed as serving a dual, but strongly interrelated role: first, knowledge creation in form of new products and production processes and, second, enhancement of the learning capacity of the firm.

Firms, especially smaller ones that lack appropriate in-house R&D facilities, have to develop and enhance their absorption capacity by other means, such as learning from customers and suppliers, interacting with other firms and organisations, and taking advantage of knowledge spillovers. These sources provide the know-why and know-how important for entrepreneurial success. Network arrangements of different kinds assist firms to take advantage of outside knowledge.

Closely related to the concept of absorption capacity, is Suarez-Villa's (2000) concept of innovative capacity, which posits that diffusion occurs through the accumulation of knowledge provided by the stock of patented inventions. These are a repository of new ideas and knowledge that have passed the scrutiny and novelty through the patent review process. Such knowledge is important as an element of the diffusion of disembodied knowledge. Also, the stock of knowledge that the innovation capacity indicator represents has potentially strong tie to diffusion within networks.

Economists use the term *knowledge spillovers* to capture the idea that knowledge created by one agent can be used by another either without compensation or with compensation less than the value of the knowledge.

Knowledge spillovers are basically externalities that flow between adjacent producers and/or users of innovation. They are especially likely to result from basic research, but they are also produced by applied R&D. This can occur in obvious ways, such as reverse engineering of products, but also in less obvious ways, such as where one firm's abandonment of a particular research line signals to other firms that the line is less productive. Knowledge spillovers may also occur when researchers leave a firm and take on a job at another firm.

Knowledge spillovers arise because, as Romer (1990) emphasised, knowledge is a partially excludable and non-rival good. *Lack of excludability* implies that knowledge producers have difficulty in fully appropriating the returns or benefits and in preventing other firms from utilising the knowledge without compensation or with compensation less than the value of knowledge. Patents and other devices, such as lead times and secrecy, are a way in which knowledge producers partially capture the benefits related to knowledge creation.

Through non-rivalry, knowledge distinguishes itself from all other inputs in the production process. *Non-rivalry* means essentially that a new piece of knowledge can be utilised may times and in many different circumstances, for example, by combining with knowledge coming from another domain. The interest of the knowledge users is, thus, best served if knowledge and innovations, once produced, are made widely available and diffused at the lowest possible cost. This implies an environment rich in knowledge spillovers best found in learning regions.

Regions are becoming focal points for knowledge generation and diffusion and learning in the new age of knowledge based economy, as they in effect become learning regions. Learning regions, as Florida (1995) points out, function as collectors and repositories of a knowledge and provide an environment or infrastructure that facilitates the flow of knowledge and learning. More generally, learning regions may be viewed as virtuous regional systems of innovation. The notion systems of innovation is conceptualised in terms of organisations in the generation, diffusion, and use of innovations and their interrelations. The learning made possible by interaction among firms and other organisations is central to the notion.

3 Networks and Network Formation

The centrality of knowledge spillovers in the process of knowledge generation and innovation is at the root of the formation of formal or informal networks. The existence of knowledge spillovers suggests that knowledge creation and innovations by a particular firm depend not only on its own efforts but also on outside efforts. Network activities assist the firm to find information and knowledge that it can not generate internally. It can – when it is part of or generates a network form of collaboration with other firms or organisations – create greater opportunities for learning, an essential prerequisite for productivity improvement, and, thus, offer a way to improve economic performance.

3.1 What is a Network?

Despite widespread use, there is no agreement on the appropriate definition of networks. The definition offered by Tijssen (1998, p. 792) captures perhaps the most important features characteristic for the network mode of organisation. He defines a network as 'an evolving mutual dependency system based on resource relationships in which their systemic character is the outcome of interactions, processes, procedures and institutionalisation. Activities within such a network involve the creation, combination, exchange, transformation, absorption and exploitation of resources within a wide range of formal and informal relationships'. He continues to characterise the network resources as "various kinds of capabilities, competencies and assets, which can be divided into 'tangibles' (e.g., codified knowledge, ...), and 'intangible' resources (skills, know-how, experience, and personal contacts)".

Network relationships generally take a long time to develop. However, once established they tend to be characterised by a high degree of interdependence, intensive communication, reciprocity, and trust. The focus of network analysis is on the interactions and relationships between the interdependent actors rather than on the actors themselves. The interaction between actors comprising a network reflects not only market relationships but also the wider social and cultural context, particularly the social rules, cultural norms, routines, and conventions (often described as institutions) through which interaction between organisations is regulated.

3.2 Different Types of Networks

The characteristics of the network vary with the type of technology and innovation, the sector of industry, and the regional or national environment. Equally varied are the types of networks. It is important to distinguish between different types of networks, such as vertical and horizontal relationships. *Vertical collaboration* occurs along the production chain for particular products, and *horizontal collaboration* occurs between partners at the same level of the production process. Vertical links are known to play a centrally important role in the innovation process. Vertical collaboration exists both within and between firms as they attempt to co-ordinate and configure their production chains. *Networks of internalised relationships* (internal coupling) incorporate R&D, design, engineering and industrial manufacturing more closely. *Networks of externalised relationships* are relational structures between independent firms that are based upon a high degree of trust (that is, expectations of honest, non-opportunistic behaviour) that takes time to develop.

The shape of externalised relationships may widely differ according to the nature of relationships between the various actors involved. There are informal and formal networks. *Formalised collaboration* between firms can take a variety of forms. It may be a *joint venture*, formed by two or more partners as a separate company on a shareholder basis. It can also be a partnership linking firms on the basis of continuing commitment to shared technological and business objectives, without equity sharing. Non-sharing agreements represent a less formalised and more flexible form of collaboration that better reflects the separate identity of the participants. In contrast to formal networks, *informal relationships*, for example, between individual researchers and between laboratories located in different institutional settings, are highly trust-based and have lower transaction costs. Transaction costs are low because decisions to trade or not to trade proprietary knowledge are made by the individual knowledgeable engineers or researchers.

3.3 Motivations for Interfirm Co-operation

Although interfirm collaboration occurs in many different forms and may reflect different motives, a number of generalisable assumptions tend to underpin them. First, there is the belief that collaboration can lead to mutual benefits that the actors involved would not achieve independently. Such benefits include *shared costs and risks* (especially important in the more expensive, speculative and risky projects in a firm's R&D portfolio), *increased scale and scope of activities,* and *improved ability to deal with the complexity and heterogeneity of multiple sources and forms of technology.*

A second assumption underlying collaboration is related to the belief that network agreements offer firms a way of ensuring a *higher degree of flexibility* in their operations compared to their alternatives. For example, collaboration can be an alternative to mergers and acquisitions, which are less easy to dissolve if necessary. The knowledge intensity of production does not necessarily imply the capacity of every firm to carry out in-house R&D. But it certainly requires firms to belong to one or several networks where R&D is performed and know-how is being created and circulated. Intangible technological knowledge is difficult to price. Network arrangements can provide a means of knowledge exchange without necessarily resorting to prices. Collaboration may allow firms to obtain access to specific tacit knowledge that may represent an important input to the generation of new products and processes.

Another assumption relates to the *time advantages* that networks tend to have over internal development and the acquisition through arm's length transactions. As product life cycles have shortened and competition has intensified, speed has become increasingly important. The time required to establish expertise or to gain market share internally is, as Porter and Fuller (1986) emphasise, likely to exceed the time required by the network mode of co-operation.

If these assumptions are correct, then there is no doubt that the network mode tends to be superior to both externalised market-governed and internalised hierarchically governed transactions in matters pertaining to the production and use of new technology. Economic instability, technological uncertainty, and the risks associated with uncertain technological trajectories have reduced the advantages of hierarchically governed transactions and make market-governed transactions a less efficient way of coping with market imperfections. The limitations of these two modes of transactions have pushed the network mode of governance to the forefront of corporate strategy in the last few decades to reduce uncertainty and to minimise transaction costs. Network co-operation has become an increasingly efficient and innovative way of organising interdependent activities in the new economy. Modern innovation competition tends to favour the network made over markets and hierarchies. Table 1 summarises some differences between these three modes (markets, hierarchies and networks) and supports the view to consider the network mode as a distinct form of economic organisation.

Table 1   Comparison of three different forms of economic organisation[a]

Characteristics	Market-governed transaction	Hierarchically governed transaction	Network mode
Normative basis	Contract property rights	Employment relationships	Complementary strengths
Communication means	Prices	Routines	Relations
Conflict resolution	Haggling Resort to courts for enforcement	Administrative fiat supervision	Norm of reciprocity Reputable concerns
Flexibility	High	Low	Medium to high
Commitment between economic actors	Low	Medium to high	Medium to high
Relations between economic actors	Independence	Hierarchical	Interdependence
Tone of climate	Precision and/or suspicion	Formal bureaucratic	Open-end mutual benefits

[a] adapted with minor changes from OECD (1992), with permission. The initial conception of this table, however, has to be attributed to Powell (1990).

Although the time is ripe for the development of a fully-fledged economic theory of networks, most studies are still couched in the framework of the Coase-Williamson theory of markets and hierarchies and make only very limited use of the concepts and notions described in the previous section. In the section that follows, we focus on networks of externalised relationships. This allows us to explore the increasingly diversified forms of interfirm relations.

4   Networks of Externalised Relationships

Interfirm collaboration is not new. What is new is the current scale, proliferation, and the fact that this mode of economic organisation has become central rather than peripheral for firms in the new economy. The increasing complexity, costs, and risks involved in innovation activities enhance the importance of interfirm collaboration and networking to reduce transaction costs.

Firms depend on a variety of external linkages in acquiring the necessary technical, scientific and organisational information and knowledge. Despite the variety there is empirical evidence that most economic activity – at least in leading industries such as electronics and automobiles – are organised around five major types of networks:

(a) *Supplier networks*, which are defined to include subcontracting, original equip-ment manufacturing, and original design manufacturing arrangements between the principal firm and its suppliers of intermediate production units.

(b) *Customer networks*, which are defined as the forward linkages of manufacturing firms with distributors, marketing channels, value-added resellers, and end users.

(c) *Technology co-operation networks*, which bear to some degree or another on technology.

(d) *R&D co-operation networks*, including university-industry relations (pre-competitive stage) pursued to gain rapid access to new scientific and technological knowledge and to benefit from economies of scale in joint R&D.

(e) *Producer networks*, which are defined to include all coproduction arrangements that enable competing producers to pool their production capacities, financial, and human resources in order to broaden their product portfolios and geographic coverage.

The first two types represent vertical and the others horizontal network relation-ships.

4.1 Supplier Networks

These networks represent a growing form of interfirm co-operation. During the past two decades, the relationship between firms and their suppliers have become increasingly complex with, in general, a tendency for major firms to develop longer term subcontracting relationships with their key suppliers and giving them greater responsibility for quality control. Subcontracting differs from the mere purchase of ready-made components and parts from suppliers in that there is an actual contract between the participating firms setting out the specifications for the order. The firm placing the order is known as the principal firm, whereas the firm carrying out the order is known as subcontractor or supplier.

Subcontracting occurs in both industrial and commercial spheres. It may cover processes and components, but also complete finished products. *Commercial subcontracting networks* involve the manufacturing of a finished product by one or more subcontractors to the principal's specification. The principal firm may be either a producer firm or a retailing-wholesaling firm. Dicken (1998) distinguishes three major types of industrial subcontracting: cost-saving subcontracting, speciality subcontracting, and complementary subcontracting. *Cost-saving subcontracting* is based on differentials in production costs between the principal and the supplier. *Speciality subcontracting* involves carrying out specialised functions for which the subcontractor has special skills and equipment, whereas *complementary sub-contracting* is adopted to cope with occasional surges in demand without expanding the principal firm's own production capacity.

4.2 Customer Networks

Many companies are paying increasing attention to customers and users as a way to obtain feedback and ideas that can help target their innovative efforts more effectively. Knowledge generated as a result of learning-by-using can be transferred into new or improved products only if the producers have close contacts with users. In turn, user firms will generally need information about new products or components. This may mean not only awareness but also quite specific inside information about new, user-value characteristics related to their specific needs. The interaction with customers is so important that it has become one of the key topics in the research on national and/or regional systems of innovation.

4.3 Technological Co-operations

New forms of interfirm agreements bearing on technology have developed alongside the traditional means of technology transfer (licensing and trade in patents) and they have often become the most important way for co-operations as well as for countries and regions to gain access to new knowledge around key technologies.

Technology co-operation networks facilitate the acquisition of product design and production technology, enable joint production and process development, and permit generic scientific knowledge and R&D to be shared. Three types of arrangements are particularly common and are discussed in OECD (1992):

(a) *Equity based agreements* that involve the use of corporate venture capital by large firms to identify innovative processes within smaller firms and to monitor the development of new technologies. This form of network permits large firms to take advantage of the continuous flow of new technologies produced by small high-tech companies.

(b) *Non-equity co-operative agreements between two firms* with the purpose of dealing with specific research problems. Here, agreements are usually of limited duration and aim at strictly defined results. Siemens and Philips are known to operate with agreements of this type.

(c) *A variety of technological agreements between firms* concerning completed technology, which include technology sharing agreements, complex two-way licensing, and cross-licensing in separate product markets.

4.4 Research and Development Co-operative Networks

R&D co-operation in isolation or in combination with other objectives represent an important objective of interfirm network arrangements in the face of rapid and radical technological change, especially in biotechnology based sectors, in information technology and in new materials. Informal networks between individual researchers and between laboratories situated in different settings (academia and business world) have a very long tradition. However, they have become increasingly formalised and have, thus, received greater institutional visibility.

University laboratories are viewed as valuable in two major respects: as an extension of research capacity and as some sort of antenna for emerging new ideas and developments in the international arena. Successful links are based on relationships founded on person-to-person familiarity, trust and reciprocity. Private sector joint ventures financed by a number of firms on a shareholder basis represent a novel development of R&D co-operation. Such firms perform their research in separate laboratories or in other research facilities established for that purpose. In general, they are staffed by personnel assigned by the partners and by personnel hired specifically for the project, and focus on generic technology in general that relates to the competitive interests of the joint venture partners. The activities, though pre-competitive in the sense of being generic, have a clear commercial focus. The European Computer Research (Centre ICL, Bull, and Siemens) is an example of this type of joint R&D venture.

4.5 Production Networks and Strategic Alliances

These networks in general are either centred on a major transnational enterprise or formed on the basis of strategic alliances between such enterprises. But they may also involve small and medium sized enterprises, particularly in knowledge-intensive sectors such as software design and biotechnology as Suarez-Villa (1998) points out. *Strategic alliances* are very different from the traditional form of cartels and other oligopolistic agreements. They concern specific times, markets, products and processes, and they do not exclude competition in all of those business activities not covered by the agreements. The motivations for strategic alliances are very specific. *Joint manufacturing agreements* are used to achieve economies of scale and to cope with excess or deficient production capacity. Such agreements tend to be successful when the relationships are structured with a clear understanding of the grand rules (especially protection of intellectual property rights) and when a positive co-operative sharing of information and knowledge is established.

Strategic alliances are clearly more difficult to manage and co-ordinate than other forms of network co-ordination, particularly when the organisations are competitors. The potential for misunderstanding and disagreement, especially between partners from different cultures, is great. Thus, many such alliances have only relatively short lives. Nevertheless, the attraction of this form of network arrangement is likely to guarantee their continued growth as a major organisational form in the new economy.

4.6 The Technological Environment to Manage the Network Arrangements

Larger companies generally participate in a series of networks (strategic alliances, subcontracting arrangements, and other forms of agreement), each with partners that have their own collaborative arrangements. The complexity of such webs would have been impossible to manage without qualitative advances in information and communication technologies in the 1990s. The growth of the Internet technologies spurred technical innovations that have produced viable economic and practical opportunities for firms to link customers, partners, and internal processes in an end-to-end business chain. Hypertext transfer protocol (HTTP) and hypertext mark-up language (HTML) allow Internet technologies to manage more complex and rich files. The user only needs a browser, which is a software program that displays HTML files on a computer screen. Today Internet technologies foster the exchange of all types of codified knowledge.

Intranets and extranets allowed the emergence of fully interactive, computer-based flexible processes of management, production, and distribution, including simultaneous co-operation between different firms and members of such firms. Intranets are local area networks (LANs) running transmission protocol-Internet protocol (TCP-IP) and supporting Internet protocols. They are playing an increasingly important role in managing the knowledge infrastructure of an organisation. Extranets provide secure limited intranet access to selected suppliers, customers, and other network partners. Because they use the same network protocol as the Internet, transmissions can be conducted on that public network. The extranet extends the intranet by creating tunnels of secure data flows using cryptography and authorisation structures. Access is available only to authorised persons, regardless of location. Thus, the extranet establishes a secure collaborative environment to manage the web of the firm's network arrangements linking core internal systems, such as materials management to complementary customers, suppliers and other network partners.

5   The Emergence of the Network Enterprise of Corporate Organisation

This section continues the discussion on co-operation but moves from interfirm co-operation to the emergence of the network form of corporate organisation, briefly termed the *network enterprise*. This form of corporate organisation, viewed by Castells (1996) as the organisational expression of the new economy, can be found in sectors other than those where innovation networks are frequent.

5.1 General Characteristics of the Network Enterprise

The notion of the network enterprise originates from studies of the Japanese corporation and its particular structure. Network enterprises are characterised by a series of distinct features. First, co-operation is viewed as a *central managerial and organisational concept*, placing joint ventures and other forms of longer term formal or informal interfirm agreements at the very centre of the enterprise. Second, this holds also true for the *internal governance of the enterprise's activities*, leading to a closer integration between R&D, design, engineering and industrial manufacturing.

Third, the application of Japanese management principles and the potential of novel information and communication technologies have given rise to a system that involves extensive use of *electronic quasi-integration* (via intra- and extranets) between the core corporation and the decentralised quasi-independent and independent production units. This makes greater internalisation of important network externalities possible as the corporation grows. In addition, it gives suppliers and other network partners greater responsibility for all aspects of component and subassembly development, design, production, and delivery in order to attain higher levels of quality.

Fourth, *new ways of managing the labour process* emphasise the capability of workers to cope with local emergencies autonomously that is developed through sharing knowledge on the shop floor and learning-by-doing. The aim of such network enterprises is to improve co-ordination of day-to-day activities by enabling users within the company and selected partners to share information and processing capacities.

In contrast to Japan, the rise of such network firms in Europe and the United States is recent and still relatively uncommon. It is directly related to qualitative advances in Internet related information technologies (intranets and extranets) in the 1990s that allowed the emergence of fully interactive, computer-based flexible processes of management, production, and distribution involving simultaneous co-operation between different forms and units of such firms.

The emergence of the network form of corporate organisation may appear in two ways: first, through the transformation of strongly centralised and hierarchical transnational co-operations into decentralised enterprises based on intracorporate networks, and, second, through the creation of genuine network enterprises via the adoption of this model for the growth and international expansion of a medium sized domestic form. Outside Japan, network enterprises have primarily appeared in Italy. Changes in large well established transnational corporations with established corporate tradition and strong organisational structures are slow, but tend to point to the elaboration of similar forms of transnational corporate organisations.

5.2 The Japanese Model of the Network Enterprise

The network enterprise is a traditional Japanese institution based upon the historical forms of the Japanese enterprise, namely the *keiretsu* type of large group structures. These forms have been used to develop stable long-term interfirm contractual relationships and have led to network types of corporate organisation.

A keiretsu is a highly diversified group of complementary enterprises that are tied together through reciprocal shareholdings, credit and trading relations, and interlocking directorships. Two basic types of keiretsu may be distinguished: horizontal keiretsu and vertical keiretsu. *Horizontal keiretsu* is a diversified industrial group organised around two key institutions: a core bank, and a general trading company, which acts as a general intermediary between suppliers and consumers. Three of the horizontal keiretsu groups (Mitsubishi, Mitsui and Sumitomo) are the heirs of the *Zaibatsu*, the giant conglomerates that led Japanese industrialisation and trade before the World War II. *Vertical keiretsu* is organised around a large parent company in a specific industry (for example, Toshiba, Sony, Toyota). Although distinctive, the two types of group are not mutually exclusive. A keiretsu formed through horizontal ties accommodates enterprises that are based on vertical integration of component suppliers and industrial subcontractors.

Following Imai (1988), Japanese network enterprises make generally use of three types of networks in their domestic operations:

(a) Wider intragroup affiliated or *horizontal keiretsu networks*, which create a form of organisation intermediate between vertically integrated firms and arm's length markets.

(b) *Vertical keiretsu networks* or supplier networks characterised by long-term and stable interfirm relationships based upon mutual obligations and comprising hundreds and even thousands of suppliers and their related subsidiaries.

(c) *R&D networks*, which consist of co-operative R&D relationships that may be separately or simultaneously intergroup, intragroup or external (for example, with universities or public research institutions).

The electronics manufacturer Toshiba, for example, is a parent company that controls R&D networks and an active supplier network with substantial numbers of satellite companies in a vertically integrated Toshiba group. Knowledge is shared and suppliers are integrated into future strategies. At the same time, Toshiba is also a member of the horizontally integrated Mitsui industrial group that is organised around the Sakura Bank and the general trading company Mitsui & Co.

5.3 Genuine Network Enterprises: The Italian Evidence

Outside Japan, the network form of organisation has emerged primarily in Italy, and here in firms with a strong regional basis in the industrial districts of northern Italy. These firms have established a wide variety of relationships, ranging from loose consortia to high levels of quasi-integration, in order to develop and market new products especially in textile industry. For small firms in market-dependent industries, economies of scale may be attained in development technology (new materials), developing joint solutions to technology diffusion (new process equipment), and marketing (export arrangements). Networking yields such economies of scale by changing the balance of facilities between the activities performed within the enterprise and those carried out externally. This is done by activities under the partial control of the individual firm but also with reliance on external collaborators. The result has been the growth of organisational forms that have some characteristics of integration combined with decentralisation.

Networking with high levels of quasi-integration can be found in firms like Benetton, as exemplified in OECD (1992). Its development displays great flexibility to adapt to changing competitive circumstances and is characterised by the combination of three elements. The first is a strong in-house capacity for design, styling, and advertising. The second is strongly decentralised manufacturing. Over 80 percent of output is produced by over 350 small and very small (quasi-) independent enterprises, most of which existed prior to their entry into the Benetton network. The third is a highly decentralised sales network with a two-tier structure. Seventy-five firms operate as agents for the group, gather orders, super-vise, and promote sales of about 4,200 shops owned by approximately 2,500 independent firms. Shops and commercial agents are connected to the Benetton Group through the same kind of relation that is characteristic for the company and its manufacturing suppliers. The parent company develops the strategy and controls the critical resources of all the autonomous firms involved in the sales network. It is worth noting that this element is novel and little known to Japanese network enterprises that use keiretsu trading companies for their marketing.

5.4 New Types of Transnational Corporations with Network-Like Forms

Transnational network corporations are still uncommon outside Japan, especially in manufacturing. But organisational changes under way point to the emergence of new types of transnational companies with network-like forms of organisation.

When demand has become unpredictable in quantity and quality, when markets have been diversified worldwide and, thus, difficult to control, and when the pace of technological change has made single-purpose production equipment obsolete, the standardised mass production system, embedded in the management methods known as Taylorism, has become too rigid and too costly for the characteristics of the new economy. New forms of work organisation and corporate management, most of them originating in Japanese firms, emerged during the 1980s in response to the above deficiencies of Taylorism. A new model of corporate organisation known as Toyotism has been widely, though hesitantly, adopted by companies and has increased flexibility and quality of production while retaining the main advantages of standardisation. Some elements of this model are well known: the just-in-time system of supplies, by which inventories are reduced substantially through the deliveries from the suppliers to the production site at the required time; total quality manufacturing that requires close co-operation between firms and their long-term subcontractors; a closer integration of R&D, design, engineering and industrial manufacture.

The move towards more decentralised structures within co-operations has also led to a major change in the approach of how human resources are managed, emphasising team work, rewards for team performance, greater autonomy of decisions on the shop floor, and a flatter management hierarchy. The stability and complement of relationships between the core firm and the suppliers' network are crucial for the shift towards a network form of organisation.

6 By Way of Conclusions

There is widespread agreement that the emergence of the knowledge-driven economy is not only changing the role of the firm but also requiring governments of nation states to reconsider their role. National governments need to recognise and take into account that their autonomy is increasingly limited by the instability and the mobility of many of the critical assets of more competitive firms.

It is becoming more and more evident that in this emerging era, the state will no longer be able to provide on its own much of the support needed to foster knowledge creation and diffusion. The retreat of the state and of many of its past functions is likely to be a painful symptom of the rise of the new economy, leaving only few basic services or entitlements untouched. Societies that seek to thrive in the knowledge-driven economy must meet this challenge head-on or else face the degradation of their knowledge potential.

A major challenge that nation state governments are facing is to balance strategic objectives concerning employment and income distribution with objectives of competitiveness and growth. The instruments available to manage the knowledge and regulatory infrastructure with which governments can counter market failure or improve the efficiency of markets are limited more than ever in their effectiveness. If massive government failures are to be avoided, governments have to realise that networks and network enterprises – and not firms in the traditional sense – have become the actual operating units in the new economy.

References and Further Readings

Castells M. (1996): *The Rise of the Network Society*, Blackwell, Oxford [UK], Malden [MA]

Cohen W. and Levinthal D. (1989): Innovation and learning: The two faces of R&D, *Economic Journal* 99 (397), 569-596

Dicken P. (1998): *Global Shift: Transforming the World Economy*, The Guilford Press, New York, London

Dosi G. (1988): Sources, procedures and microeconomic effects of innovation, *The Journal of Economic Literature*, 26 (3), 1120-1126

Dosi G., Freeman C., Nelson R.R., Silverberg G. and Soete L. (eds.) (1988): *Technical Change and Economic Theory*, Pinter, London

Fischer M.M. and Fröhlich J. (eds.) (2001): *Knowledge, Complexity and Innovation Systems*, Springer, Berlin, Heidelberg, New York

Fischer M.M., Revilla-Diez J. and Snickars F. (2001): *Metropolitan Innovation Systems: Theory and Evidence from Three Metropolitan Regions in Europe*, Springer, Berlin, Heidelberg, New York

Fischer M.M., Suarez-Villa L. and Steiner M. (eds.) (1999): *Innovation, Networks and Localities*, Springer, Berlin, Heidelberg, New York

Florida R. (1995): Toward the learning region, *Futures* 27 (5), 527-536

Imai K. (1988): International corporate networks: A Japanese perspective, Project Prométhée Perspectives, June

Jaffe A.B., Trajtenberg M. and Henderson R. (1993): Geographic localization of knowledge spillovers as evidenced by patent citations, *Quarterly Journal of Economics* 108 (3), 577-598

Lundvall B.-Å. (ed.) (1992): *National Systems of Innovation: Towards a Theory of Innovation and Interactive Learning*, Pinter, London

Malecki E.J. (1997): *Technology & Economic Development. The Dynamics of Local, Regional and National Competitiveness*, Longman, Harlow, Essex

Nelson R.R. and Winter S.G. (1982): *An Evolutionary Theory of Economic Change*, Belknap Press of Harvard University Press, Cambridge [MA]

Nonaka I. and Takeuchi H. (1995): *The Knowledge-Creating Company. How Japanese Companies Create the Dynamics of Innovation*, Oxford University Press, New York, Oxford [UK]

OECD (1997): *Industrial Competitiveness in the Knowledge-Based Economy: The Role of Governments*, Organisation for Economic Co-operation and Development, Paris

OECD (1992): *Technology and the Economy: The Key Relationships*, Organisation for Economic Co-operation and Development, Paris

Polanyi M. (1967): *The Tacit Dimension*, Doubleday Anchor, New York

Porter M.E. and Fuller M.B. (1986): Coalitions and global strategy. In: Porter M.E. (ed.) *Competition in Global Industries*, Harvard Business School Press, Boston, pp. 315-344

Powell W.W. (1990): Neither market nor hierarchy: Network forms of organisations, *Research in Organisational Behaviour* 12, 295-336

Romer P.M. (1990): Endogenous technological change, *Journal of Political Economy* 98, 71-102

Saviotti P.P. (1998): On the dynamics of appropriability, of tacit and of codified knowledge, *Research Policy* 26 (7/8), 843-856

Suarez-Villa L. (2000): *Invention and the Rise of Technocapitalism*, Rowman & Littlefield, Lanham

Suarez-Villa L. (1998): The structures of co-operation: Downscaling, outsourcing and the networked alliance, *Small Business Economics* 10, 5-16

Tijssen R.J.W. (1998): Quantitative assessment of large heterogeneous R&D networks: The case of process engineering in the Netherlands, *Research Policy* 26 (7/8), 791-809

Appendix

Absorptive capacity: The ability of a firm to integrate new pieces of knowledge into its own knowledge stock.

Codified knowledge: Knowledge recorded in codes that can be stored for access at other times and locations.

Disembodied knowledge: Knowledge that is disentangled from the product or service itself, for example through licenses or patents.

Dissemination: The conversion from codified to codified knowledge by the owners sharing it with one another.

Embodied knowledge: Knowledge that is captured in products or services.

Externalisation: The conversion from tacit knowledge to codified knowledge, generally by articulating the tacit knowledge and turning it into explicit form.

Horizontal keiretsu: Diversified industrial groups in the contemporary Japanese economy organised around a core bank and a general trading company (for example, Mitsubishi).

Innovation – the object view: A novel product or practice which is made available for application, with commercial success. Characteristically, different degrees of novelty may be distinguished, depending on the context: an innovation can be new to an organisation, new to the market, new to a region or country, or globally new.

Innovation – the process view: Process of using, applying and transforming scientific and technical knowledge in the solution of practical problems. Today, the innovation process may be characterised by continuing interaction and feedback.

Internalisation: The conversion from explicit (codified) knowledge back to tacit knowledge, closely related to learning-by-doing.

Invention: A new idea that has not yet been commercialised.

Knowledge: The definition of this concept varies. Some define knowledge in terms of information theory and believe that knowledge is information oriented (the knowledge-object camp). Others view knowledge as a process and see knowledge embodied in the organisation's employees and business processes (knowledge process camp).

Knowledge base: Contains knowledge in all its forms, from simple to routine procedures of everyday production activities to shared practices and habits of thought.

Knowledge conversion: Addresses the problem of pulling knowledge created by individuals into the business organisation's knowledge base where it can then be managed and shared. There are four phases of knowledge conversion: Externalisation, socialisation, internalisation, and dissemination.

Knowledge spillovers: Leakage of knowledge from the generating agent to other agents (without compensation or with compensation less than the value of the knowledge).

Knowledge-creating company: A company based on the organisational interaction between codified and tacit knowledge.

Knowledge-driven economy: An economy in which knowledge creation and diffusion are in some sense qualitatively more important than ever before. The dynamics of the economy are coming to rest less on investment in physical capital and more and more on investments in knowledge creation.

Learning: The process of acquiring and applying new information and skills. It is recognised as a critical component in the development of continuous innovation for organisations.

Learning organisation: Organisation that knows to use knowledge and gives its employees the opportunity and the tools to develop and apply knowledge.

Network: A distinct form of interfirm organisation that operates alongside and in combination with markets and 'hierarchies' (that is large centralised forms of corporate organisation built on vertical and horizontal integration).

Network enterprise: Network form of corporate organisation that is viewed as the organisational expression of the new economy.

Socialisation: The conversion from tacit knowledge to tacit knowledge through sharing experiences, imitation etc.

Tacit knowledge: Knowledge that is ill-defined, uncodified and possibly uncodifiable, and differs from person to person.

Vertical keiretsu: Diversified industrial groups in the contemporary Japanese economy organised around a large parent company in a specific industry (for example, Toyota, Sony).

6 The Innovation Process and Network Activities of Manufacturing Firms

This chapter continues to contribute to our understanding of both the innovation process and the process of network formation by emphasising that the interactive nature of the innovation process has broken down the distinction between innovation and diffusion so that the creation of knowledge and its assimilation via networks are part of a single process. The discussion is enriched with empirical evidence from a comprehensive questionnaire of manufacturing firms in the metropolitan region of Vienna. The study confirms the vital importance of customer and user producer linkages and co-operation in the pre-competitive stage of the innovation process.

1 Introduction

Manufacturing firms in Europe have come under increasing pressure in recent years. This pressure arises from three major phenomena and processes that are affecting the entrepreneurial environment: *first*, the transition from internationalisation to globalisation, accompanied by a process of global concentration in a number of industries, *second*, the establishment of the Single European Market and the prospects of the Economic and Monetary Union, and *third*, the opening of the Iron Curtain and increasing competition from the newly developing market economies in Eastern Europe. Firms may react in different ways to meet these challenges, but there is wide agreement that new technologies, along with novel forms of work organisation and management, will play a crucial role in enabling firms to respond successfully to rapidly changing market conditions and remain competitive in an increasingly European or even global economic environment.

This contribution focuses on innovation and network activities and identifies the reasons why we need a better understanding of both the innovation process and the process of network formation. This response is largely conceptual, based on the body of evolutionary theory of economic change that comprises a rich environment of learning and interaction, which are the two central elements in the current understanding of the process of innovation (Nelson and Winter 1982; Dosi 1988; Lundvall 1988, 1992; Suarez-Villa 1989). Some empirical evidence will be provided from a survey carried out in the metropolitan region of Vienna.

The contribution is organised as follows. The next section provides a basic account of the key elements of the analysis: technology, codified and uncodified knowledge, and innovation. In Section 3 we then describe the nature of the innovation process on the basis of current thinking and understanding which emphasise three major elements: the role of design in the wider sense, learning that allows firms to create dynamic advantages, and interaction, both internal to firms and external, i.e., with other firms and institutions.

Section 4 moves on to the diffusion of disembodied knowledge. Special attention is given to the notions of knowledge spillovers and the absorption capacity of a firm. Both play a central role in achieving a deeper understanding of the external network activities of firms, which is the focus of Section 5. The line of reasoning starts with a characterisation of the network mode of organisation that provides the necessary relations to use outside knowledge. It then continues to discuss in which circumstances this mode is superior to market transactions and vertical integration, the two forms of organisation previously recognised by economic theory, and finally points to the diversity and localised nature of networks. Section 6 then presents some empirical evidence of innovation and network activities of manufacturing firms in the metropolitan region of Vienna. The concluding section summarises some of the major findings of the discussion.

2   Technology, Knowledge and Innovation

Innovation – in the form of advancing technology – provides the principal source of change for firms, regions and nations. It is, however, a complex concept with many meanings. For the purpose of this contribution, it is important to provide at this juncture working definitions of technology, knowledge and innovation.

We will begin by defining *technology* in accordance with (Mansfield et al. 1982) as consisting of a pool or set of knowledge. It is important to distinguish knowledge from information. Information may be interpreted as factual (Saviotti 1988), while knowledge establishes generalisations and correlations between variables (Andersson 1985). Particular pieces of information can be understood merely in the context of a given type of knowledge, for example a theory. New knowledge creates new information and this information can be understood and used only by those who possess the new knowledge. In this sense knowledge has a *retrieval/interpretative* and not only a *correlational* function (Saviotti 1998).

Knowledge has some further outstanding characteristics that it is important to mention. Firstly, knowledge is *cumulative* (Teece 1981; Nelson and Winter 1982), which implies path-dependence and the possibility of creating barriers, since established participants – in given technologies – accumulate a differential advantage with respect to potential entrants. Knowledge in firms also has a *collective character*. This means that knowledge is not simply the sum of the pieces embodied in the individual workers of the firm (Saviotti 1998). In this sense, the knowledge base of a firm may be defined as the collective knowledge that a firm uses to produce its output.

The knowledge base contains knowledge in all its forms, from simple and routine procedures of everyday life to the methods of organisation and management, from the machinery (i.e., embodied knowledge) to the scientific concepts, methods and theories that enable newer inventions. In most cases, a piece of knowledge can be located somewhere in a range between the completely tacit and completely codified. Knowledge is always at least partly tacit in the minds of those who create it. The process of codification is necessary because knowledge production is a collective undertaking that requires communication. The transmitter and the receiver have to know the code to be able to communicate. The codification process for a given subject amounts to the gradual convergence of the scientific community and of other users on common standardised definitions and concepts, on common contents and theories. The degree of codification differs for different types of knowledge at a given time. Knowledge closer to the frontier, and therefore more recent, is likely to be more tacit than knowledge which is already established (Saviotti 1998).

Codified knowledge is that form of knowledge which is in some way tangible – usually found in print form, such as scientific papers and patent applications. Much knowledge is codified and publicly accessible. But much of the essential knowledge – especially the newer parts that constitute the frontier – resides in tacit form in the minds of experienced individual researchers or engineers. This person-embodied knowledge is generally difficult to transfer, and is often only shared by colleagues if they know the code through common practice. On the one hand, a given type of knowledge may become more codified as it matures, on the other, the act of embodying it into specific goods and services may reintroduce some 'tacitness'.

Traditionally, knowledge was viewed as a public good, because it is not possible for a producer of knowledge to prevent its use by economic agents who do not pay anything in exchange for it. But even a completely codified piece of knowledge can not be utilised at zero cost by everyone. Only agents who know the code can use the piece of knowledge at zero imitation cost. Others – if they realise the economic value of a given piece of knowledge – have to learn the code before being able to retrieve and imitate. Tacit knowledge is an important element of the knowledge that firms require for innovation. Such knowledge is generated in different ways, generally described as mechanisms or modes of learning. Such mechanisms or modes vary in dependence on the type of knowledge and on the institutional setting in which learning takes place.

Commercial products and production processes represent various combinations of pieces of knowledge, codified and tacit, in a specific technology set. *Innovation* is generally defined as the activities of developing and commercialising new products and processes (see e.g. Hall 1986). These innovation activities are of two major types: fundamental, which involves the creation and utilisation of a piece of new scientific, technological or organisational knowledge; and incremental, which concerns product or process improvements based on existing knowledge (Freeman 1986). The partly tacit character of knowledge is likely to be responsible for the importance that localised networks of personal contacts have for the innovation activities of firms in some metropolitan regions. The diffusion of innovations within and between firms and industries over time and space represents technological change.

3   The Interactive Character of the Innovation Process

For a long time, thinking about technological change and innovation was dominated by linear models – in the 1950s and 1960s by the technology-push and then the need-pull model. In the former, the development, production and marketing of new technology was assumed to follow a well defined time sequence which began with basic and applied research activities, involved a product development stage, and then led to production and possibly commercialisation. In the second model, this linear process emphasised demand and markets as the source of ideas for R&D activities. Despite the appealing logic of such conceptualisations, these models came under increasing attack, due in particular to the apparent disorderliness of the innovation process occurring in the post Fordist era.

Current thinking about the innovation process emphasises the tacit and non-codifiable nature of technology, the importance of learning-by-doing and learning-by-using, and the cumulative nature of learning. Learning is now widely accepted as a central element in the process of innovation. Learning allows firms to create dynamic advantages so that the force of imitation is outrun by the pace of innovation. Since innovation reflects learning as much as it does novelty, and since personal contacts are crucial for transferring pieces of tacit knowledge, the partly tacit character of scientific and technological knowledge is responsible for the central importance of interactions in the innovation process.

In line with this view, linear models of the innovation process have been supplanted by interactive models of innovation. These models stress the feedback effects between upstream (technology related) and downstream (market related) phases of the innovation process, the many interactions of innovation related activities, both within firms and in network agreements between them, and the central role of industrial design (in its widest sense) in the innovation process. Broadly speaking design includes two dimensions (Kline and Rosenberg 1986): 'initiating design', which reflects invention, and 'analytical design', which is the study of new combinations of existing products and components, or rearrangement of processes.

Interactive models portray the innovation process as a set of activities that are linked to one another through complex feedback loops. The process may be visualised as a chain, starting with the perception of a new market opportunity and/or a new invention based on novel pieces of scientific and/or technological knowledge (i.e., initiating and/or analytical design); followed by detailed design and testing, redesign and production, and distribution and marketing. Initiating and analytical design are crucial for knowledge production in order to create inventions and innovations, while redesign is important for their ultimate success. Problems arising during the processes of designing and testing new products and production processes often require links to science and especially engineering disciplines in academia.

These models recognise interaction as a central element in the process of technological innovation. Two types of interactions can occur. The first concerns interaction processes within a corporation, i.e., intrafirm networking, such as loops that link R&D with engineering and production, and loops that link different groups within R&D. These links may be complemented by interfirm networking, the second type of interaction, involving other firms and institutions of the wider science and technology environment in which the firm operates.

4 Technology Diffusion, Absorption Capacity and Knowledge Spillovers

Recognition of the interactive nature of the innovation process has resulted in the break down of the earlier distinction between innovation and diffusion. The creation of knowledge and its assimilation are part of a single process. Firms need to absorb, create and exchange knowledge interdependently. In other words, innovation and diffusion usually emerge as a result of an interactive and collective process within a web of personal and institutional connections which evolve over time.

Knowledge transfer may occur through disembodied or equipment-embodied diffusion. The latter is the process by which innovations spread in the economy through the purchase of technology-intensive machinery, such as computer-assisted equipment, components and other equipment. Disembodied technology diffusion refers to the process where technology and knowledge spread through other channels not embodied in machinery (OECD 1992). This type of knowledge transfer may occur via descriptions of new products or production processes found in catalogues, publications or patent applications, but also via seminars and conferences, and R&D personnel turnover. It can also be the byproduct of mergers and acquisitions, joint ventures or other forms of interfirm co-operation.

Two notions are central to an understanding of disembodied technology diffusion: the first is that of absorption capacity and the second that of knowledge spillovers. The *absorption capacity* of firms and research institutions refers to the ability to learn, assimilate and use knowledge developed elsewhere through a process that involves substantial investments, especially of an intangible nature (Cohen and Levinthal 1989). This capacity depends crucially on the learning experience, which in turn may be enhanced by in-house R&D activities. The concept of absorption capacity implies that in order to have access a piece of knowledge developed elsewhere, it is necessary to have done R&D on something similar (Saviotti 1998). Thus, R&D may be viewed as serving a dual, but strongly interrelated role: firstly, developing new products and production processes, and secondly, enhancing the learning capacity.

The degree of importance of R&D for the development of a firm's absorption capacity largely depends on the pace of advance in the area of technology concerned and the characteristics of outside knowledge (i.e., the degree of codification and the degree to which the knowledge can be appropriated). The faster the pace of advance in the field, the lower the degree of codification, the higher the degree of 'appropriability' and the greater the effort needed to keep up with the developments. In general, the more tacit a specific piece of knowledge, the greater the time and effort required to learn the code and to transform the knowledge into a form which is firm specific and commercially relevant.

Firms, especially smaller firms, that lack appropriate in-house R&D facilities have to develop and enhance their absorption capacity by means of other sources, such as learning from customers and suppliers, by interacting with other firms and taking advantage of knowledge spillovers from other firms and industries (Lundvall 1988). These sources provide the know-why, know-how, know-who, know-when and know-what important for entrepreneurial success (Johannisson 1991; Malecki 1997). Network arrangements of different kinds provide a firm the assistance necessary to take advantage of outside knowledge.

The diffusion of disembodied knowledge originates in the externalities which characterise the innovation process and knowledge spillovers that occur when the firm developing a piece of new knowledge can not fully appropriate the results of knowledge creation. The degree of appropriability differs for different types of knowledge at a given time. Appropriability is expected to fall systematically during the maturation of a technology as the degree of codification and the number of economic agents knowing the code increase.

Knowledge spillovers arise because knowledge and innovation are a partially excludable and non-rivalrous good (Romer 1990). Lack of excludability implies that knowledge producers have difficulty in fully appropriating the returns or benefits and preventing other firms from utilising the knowledge without compensation (Teece 1986). Patents and other devices, such as lead times and secrecy, are a way for knowledge producers to partially capture the benefits related to knowledge creation. It is important to recognise that even a completely codified piece of knowledge can not be utilised at zero cost by everyone. Only those economic agents who know the code are able to do so (Saviotti 1998).

Non-rivalry means essentially that a new piece of knowledge can be utilised many times and in many different circumstances, for example by combining with knowledge coming from another domain. The interest of the knowledge users is thus best served if innovations, once produced, are widely available and diffused at the lowest possible cost. This implies an environment rich in knowledge spillovers (OECD 1992).

The appropriability characteristics of particular technologies suggest that knowledge generation by a particular firm depends not only upon in-house R&D activities, but also on outside efforts and – more generally – on the scientific and technological knowledge pool on which it can draw. With the interactive model of the innovation process in mind, innovation and diffusion can be seen to be closely interlinked. Technology innovation leads to diffusion of knowledge that in turn affects the level of innovative activity in a firm.

5 Networks and Network Formation

In recent years, new forms of interfirm agreements bearing on technology have developed alongside the traditional means of technology transfer – licensing and trade in patents – and they often have become the most important way for firms, regions and countries to gain access to new knowledge and key technologies. The network form of governance can overcome market imperfections as well as the rigidities of vertically integrated hierarchies. The limitations of these two modes of transaction in the context of knowledge and innovation diffusion have pushed interfirm agreements to the forefront of corporate strategy in the last few decades (Chesnais 1988).

There are many definitions of *innovation networks* (see De Bresson and Amesse 1991; Freeman 1991). But the one offered by Tijssen (1998, p. 792) captures the most important features of the network mode. He suggests defining a network as 'an evolving mutual dependency system based on resource relationships in which their systemic character is the outcome of interactions, processes, procedures and institutionalisation. Activities within such a network involve the creation, combination, exchange, transformation, absorption and ex-ploitation of resources within a wide range of formal and informal relationships.' In a network mode of resource allocation, transactions occur neither through discrete exchanges nor by administrative fiat, but through networks of individuals or institutions, engaged in reciprocal, preferential and supportive actions (Powell 1990).

Networks show a considerable range and variety in content, which differs according to specific circumstances. Its nature will be shaped by the objectives for which network linkages are formed. For example, they may focus on a single point of the R&D-to-commercialisation process or may cover the whole innovation process. The content and shape of a network will also differ according to the nature of relationships and linkages between the various actors involved (see Chesnais 1988). At the one end of the spectrum lie highly formalised relationships. The formal structure may consist of regulations, contracts and rules that link actors and activities with varying degrees of constraint. At the other end are the network relations of a mainly informal nature, linking actors through open chains. Such relations are very hard to measure (Freeman 1991). Whenever interfirm transactions tend to be small in scale, variable and unpredictable in nature, requiring face-to-face contact, then network formation will focus on the close proximity of the partners involved (Storper 1997).

For firms, networks represent a response to quite specific circumstances. Where complementarity is a prerequisite for successful innovation, network agreements may be formed in response to specific proprietary tacit knowledge. The exchange of such complementary assets can take place only through very close contacts and personalised, and generally localised, relationships (OECD 1992). When tech-nology is moving rapidly, flexibility and reversibility along with risk sharing represent another reason for preferring a network mode. Interfirm agreements are easier to dissolve than internal developments or mergers. The network mode provides a far higher degree of flexibility (OECD 1992). Porter and Fuller (1986) stress speed as being among the advantages that networks have over acquisition or internal development through arm's length relationships. This advantage is becoming increasingly important as product life cycles have shortened and competition has intensified. High R&D cost may be another distinct reason for networking and can force management, especially in smaller firms, to pool resources with other firms, in some cases even with competitors (OECD 1992).

6   Innovation and Network Activities in the Metropolitan Region of Vienna

Any empirical study of innovation and network activities requires primary data collection and postal or interview based surveys which take the individual manufacturing firm as the unit of analysis. We decided that a postal survey of firms was the most appropriate methodological tool for eliciting basic quantitative data. The questionnaire used underwent several rounds of development and revision within the framework of an international project on Regional Innovation Potential and Innovative Networks in Metropolitan Regions. It was finally conducted from September 4 to December 15, 1997, in the metropolitan region of Vienna (i.e., the city of Vienna and related communities). The key questions covered the organisational structure, product and process mix, as well as the nature and extent of innovation and network activities. The questionnaire was sent to the 908 manufacturing firms with at least 20 employees, identified by the Firm and Product Database Register (1995) organised by the Department for Systems Research at the Austrian Research Centers Seibersdorf. Of these firms, 204 returned the completed questionnaire, representing a response rate of approximately 22.5 percent. This rate is relatively low, but statistically still acceptable. Anecdotal evidence indicates that industrialists are receiving postal surveys in ever increasing numbers and this inevitably has an effect on response rates.

Table 1 presents the responses broken down into seven industrial sectors (using the standard NACE classification on the basis of information such as product description, provided by the firms) and four size classes (measured in terms of employment). The sample can be seen to broadly reflect the overall structure of the total population. The slightly lower response rate from small local manufacturing units was expected and could be attributed to the fact that such firms are less likely to undertake any kind of formal R&D activity, as they tend to lack the necessary resources, and therefore display a tendency to dismiss the questionnaire as irrelevant to their circumstances. This is a general problem and not specific to this study. A telephone based survey of a small subsample of 90 non-respondents, however, indicated that the problem was not significant.

Table 1   Response patterns and representativeness of responding manufacturers

	Total number of registered firms 1995		Number of responding firms 1997		Represent-ativeness ratio [a]
Industry sector					
Textiles & clothing	72	(7.93 %)	13	(6.37 %)	18.05 %
Food industry	112	(12.33 %)	24	(11.76 %)	21.43 %
Wood, paper & printing	198	(21.81 %)	49	(24.02 %)	24.75 %
Chemicals, plastics & rubber	185	(20.37 %)	38	(18.63 %)	20.54 %
Electrical and optical equip.	115	(12.67 %)	28	(13.73 %)	24.35 %
Basic metals and metal prods.	108	(11.89 %)	24	(11.76 %)	22.22 %
Machinery & transport	118	(13.00 %)	28	(13.73 %)	23.73 %
Total	908	(100.00 %)	204	(100.00 %)	22.47 %
Employment size					
≤ 49	396	(43.61 %)	88	(43.14 %)	22.22 %
50 – 99	225	(24.78 %)	49	(24.02 %)	21.78 %
100 – 499	232	(25.55 %)	54	(26.47 %)	23.28 %
≥500	55	(6.06 %)	13	(6.37 %)	23.64 %
Total	908	(100.00 %)	204	(100.00 %)	22.47 %

[a] number of responding manufacturing firms divided by total number of registered firms multiplied by 100. Source: Innovation Survey 1997, data compiled by Vera Mayer

The majority of firms surveyed were very small (67.2 percent with less than 100 employees, compared to 68.4 percent of the identified population), and many of these (49.6 percent of those with a known starting year) have been in business since 1970. In terms of organisational status, 111 firms (55.0 percent) were independent, the remainder operated within a wider parent company group as a main plant (36.1 percent) or as a branch plant (8.9 percent).

Table 2 shows a brief profile of the surveyed firms utilising five indicators. The first three attempt to capture the resources to which the manufacturing firms have access for the purposes of innovation:

- the presence of continuous on-site R&D facilities,
- R&D employment in terms of the R&D personnel ratio,
- R&D expenditure in terms of the R&D expenditure intensity (as percentage of sales turnover).

Another set of two indicators focuses on innovation activities or outcomes and includes:

- the actual introduction of new products (averaged over 1994-1996) per 1,000 employees, i.e., the product innovation rate,
- the share of turnover accounted for by new or improved products (averaged over 1994-1996).

The second of these measures is an indicator favoured by many of management experts as a measure of a firm's innovativeness and is a widely accepted measure in the benchmarking literature (see, for example, Zairi 1992). It relates product innovations to economic activity. It is accepted that the definition of what constitutes a new or improved product is problematic and this has to be taken into account when considering the figures given in Table 2 In some industry sectors, such as the food industry and textiles & clothing, new and especially improved products may appear rapidly, while in others four or five year development cycles may be the norm. In sectors such as machinery & transport, for example, very long leading times are still the case.

Table 2   Selected characteristics of surveyed firms (1994 – 1996)

	Firms with continuous on-site R&D 1997	R&D personnel ratio [a]	R&D expend. intensity	Innovation rate [b]	% of turn-over by product innovation
Industry sector	c				
Textiles & clothing	2 (15.38 %)	17.76	5.84	62.53	0.13
Food industry	3 (12.50 %)	28.18	1.76	34.02	0.28
Wood, paper & printing	4 (8.16 %)	11.50	1.55	27.71	0.04
Chemicals	5 (13.16 %)	53.29	6.39	41.84	0.19
Electrical & optical equipment	7 (25.00 %)	250.41	16.05	6.15	0.51
Basic metals & metal products	2 (8.33 %)	26.18	2.30	11.99	0.53
Machinery & transport	7 (25.00 %)	25.50	5.21	4.01	0.50
Employment size					
≤ 49	7 (7.95 %)	51.74	2.84	128.12	0.13
50 – 99	7 (14.29 %)	30.54	3.18	86.83	0.16
100 – 499	11 (20.37 %)	32.18	4.35	6.51	0.32
≥ 500	5 (38.46 %)	142.59	10.05	2.46	0.44
Production size					
Custom production	11 (12.09 %)	39.03	4.56	27.75	0.26
Batch production	6 (10.71 %)	176.21	11.40	15.73	0.42
Custom & batch production	1 (12.50 %)	37.75	2.92	34.82	0.12
Mass production	10 (29.41 %)	66.64	6.81	6.96	0.27

[a] per 1,000 employees; [b] number of new products per 1,000 employees; [c] percentage of all firms of the corresponding raw category. Source: Innovation Survey 1997, data compiled by Vera Mayer

Following Malecki and Veldhoen (1993), we classified firms as innovative, if they met the following criterion: namely, that product innovations introduced during the past three years comprised more than 20 percent of the firm's yearly turnover. Defined in this way, only 50 (26.5 percent) of the firms were innovative. 64.0 percent of these has fewer than 100 employees, and 16 had under 50 employees. The sectoral distribution indicates a predominance of innovative firms in the category: electrical & optical equipment (ÖNACE 30-33; 11 firms), machinery &

transport (ÖNACE 29, 34-35; 11 firms) and basic metals & metal products (ÖNACE 27-28; 3 firms). These three sectors account for 50 percent of all the innovative firms. Of the non-innovative firms, 45.3 percent are engaged primarily in custom production, 26.6 percent in batch production and another 5.0 percent in custom & batch production. This suggests that flexible production, particularly of custom products for individual customers, is the norm rather than the exception among the firms surveyed, whether or not the concept of 'new/improved' products is appropriate.

R&D may be misleading or at least incomplete as an indicator of technological capability, because it does not include network activities, learning, informal R&D and other means of enhancing a firm's knowledge base (Malecki 1997). The performance of a firm may be best viewed as a product of the interplay between in-house R&D efforts to innovate and external innovation networks for knowledge transfer. The knowledge needed to compete comes most often from customers, suppliers (manufacturing and producer service suppliers) and from other firms and institutions. The innovativeness supported by regional interfirm networks not only supports existing firms, but also offers opportunities to open up new businesses in order to serve newly identified markets. The importance of networks and of innovative niches in sparking innovation applies in both high-technology industries and in traditional sectors.

Network activities of manufacturing firms in the Vienna metropolitan region are organised around five types of networks:

- *customer networks* which are defined as the forward linkages of manufacturing firms with distributors, marketing channels, value-added resellers and end users,

- *manufacturing supplier networks* which include subcontracting, arrangements between a client (the focal manufacturing firm) and the manufacturing suppliers of intermediate production inputs,

- *producer service supplier networks* which include arrangements between a client (the focal manufacturing firm) and its producer service partners (esp. computer and related service firms, technical consultants, business and management consultants, market research and advertising),

- *producer networks* which include all co-production arrangements (bearing to some degree or another on technology) that enable competing producers to pool their production capacities, financial and human resources in order to broaden their product portfolios and geographic coverage,

- *co-operation with research institutions/departments of universities* (pre-competitive stage) pursued to gain rapid access to new scientific and technological knowledge and to benefit from economies of scale in joint R&D.

Firms pursue such co-operative arrangements in order to tap into sources of know-how located outside the boundaries of the firm, to gain fast access to new technologies or new markets, to benefit from economies of scale in joint R&D and/or production, and to share the risks for activities that are beyond the scope or capabilities of a single firm. The picture which emerges from the evidence of the current study is that there exists a maze of different networks. They range from highly formalised to informal network relations, from highly specialised and rather narrow networks to looser and much wider networks such as, for example, technical alliances involving firms as corporate entities, from networks focusing on the pre-competitive stage of the innovation process to those involving the competitive stage.

Table 3 provides some empirical evidence on the five types of networks described above, from the point of view of the focal manufacturing firm, and highlights the fact that:

- Co-operation in the pre-competitive stage (i.e., in the early stages) of the innovation process is generally more common than in the competitive stage. External information tends to be particularly relevant during the early stages of the innovation process when perception of problems and evaluations of technological possibilities take place.

- Customer and user-producer (i.e., manufacturing and producer service supplier) relationships are much more frequent than horizontal forms of co-operation such as producer networks and research institution-industry linkages. Customer networks represent the most frequent form of interfirm co-operation – activities with customers and suppliers constituting 35.3 percent of all such activities. Manufacturing and producer service suppliers have strong incentives to establish close relationships with user firms and even monitor some aspects of their activity. Knowledge produced as a result of learning-by-using can only be transformed into new products if the producers have direct contact with users. In turn, user firms will generally need information about new products or components. This may not only mean awareness, but also quite specific inside information about how new, user-value characteristics relate to their specific needs.

- 37.7 percent of the manufacturing firms are integrated into customer networks, 27.9 percent into manufacturing supplier networks, 46.6 percent into producer service supplier networks, and only 18.6 percent have set up co-operative relations with research institutions and/or departments[1] of universities, despite the active promotion of university-industry programmes in Austria.

- The data clearly suggest that the significance of metropolitan co-operation between firms should not be overestimated. Spatial proximity seems to be one

---

[1]   Departments are defined as subdivisions of institutes.

Table 3 Network activities of manufacturing firms

		Customer networks		Manufacturing supplier networks		Producer service supplier networks		Producer networks		Co-operation with research institutions	
			c		c		c		c		c
Pre-competitive stage											
Information exchange	a	199		135		165		66		61	
	b	64	(26.1 %)	45	(23.0 %)	63	(34.5 %)	27	(30.3 %)	25	(32.8 %)
Identification of new ideas	a	190		122		148		64		57	
	b	57	(25.8 %)	39	(24.6 %)	57	(34.5 %)	25	(28.1 %)	20	(31.6 %)
Research and Development	a	179		118		148		49		56	
	b	55	(25.7 %)	37	(23.7 %)	56	(34.5 %)	20	(26.5 %)	22	(30.4 %)
Competitive stage											
Prototype development	a	175		108		96		37		47	
	b	53	(24.6 %)	34	(23.1 %)	36	(32.3 %)	16	(27.0 %)	20	(31.9 %)
Pilot projects	a	167		97		101		28		47	
	b	51	(25.1 %)	30	(24.7 %)	41	(34.7 %)	12	(32.1 %)	20	(29.8 %)
Market introduction	a	183		82		105		49		19	
	b	56	(26.2 %)	25	(25.6 %)	38	(34.3 %)	20	(22.4 %)	9	(31.6 %)

[a] number of such network activities of the manufacturing firms (with all regions); [b] number of manufacturing firms with such network activities (with all regions); [c] share of such network activities with a focus on the metropolitan region of Vienna. Source: Innovation Survey 1997, data compiled by Walter Rohn

criterion, but not a decisive one for innovation-oriented and even personal relationships. The building up and fostering of mutual trust is possible without the precondition of spatial proximity.

As in other studies (see, for example, Meyer-Krahmer 1985) three clusters of manufacturing firms may be distinguished. The first cluster, characterised by a high outward orientation, frequently utilises the whole range of possibilities in obtaining external knowledge. Firms in this cluster share widespread network activities in both the pre-competitive and the competitive stage of the innovation process, also with research institutions. Spatial proximity to the co-operation partners is irrelevant. Competence and excellence tends to be the decisive criterion. The second cluster of firms is characterised by medium outward orientation and seems to rely more on in-house problem solving strategies. Such firms tend to have regular contacts with customers and suppliers. Linkages with research institutions and universities are less common. Geographic proximity to co-operation partners is less important. The third cluster relies almost entirely on in-house problem solving techniques. It includes less innovative firms with less complex products and highly specialised firms that operate in small market niches. Even though the latter are quite innovative, few have network activities in the competitive stage of the innovation process.

7   Summary and Conclusions

Technological innovations represent various combinations of knowledge in a specific technology set. The processes through which innovations emerge are extremely complex. They have to do with knowledge generation and spillovers, as well as with the translation into new products and new production processes. The path followed by this translation is by no means linear, i.e., going from basic to applied research and then to the development and implementation stages of new production processes and new products. Innovation processes have an evolutionary character; they are path dependent and develop over time. They are, moreover, characterised by complex feedback mechanisms and interactive relations involving research, technology, production and the market.

Interactions occur within firms, i.e., between different individuals or departments, and between firms and customers, between different firms (e.g. firms and their suppliers), or between firms and research institutions. The survey has provided broad empirical evidence that such interfirm co-operations and networks do not take place only in the pre-competitive stage of the innovation process, but also in the competitive stage. Forward linkages (customer networks) and backward linkages (supplier networks) constitute the most important types in the Vienna metropolitan region.

The character of the network to which a firm belongs has a bearing on the type of information and knowledge to which it gives the firm access. The amounts of information and knowledge dispersed may vary between networks, and here the connectivity of the constituent parts of the network matters. The ability of small and medium sized firms to link up in networks between themselves and also with large corporations is dependent on the availability of new information and communication technologies, once the network's horizon becomes international or even global. The complexity of the web of subcontracting and co-operation agreements would be simply impossible to manage without the development of computer networks, more specifically, without powerful microprocessors installed in desktop computers linked via digitally switched telecommunication networks.

Network forms and activities are essential to the competitiveness of small and large firms. But we still know too little about how they operate. In particular, we need to know more about the variety of organisational forms, about trust and power relationships in networks. Such issues are difficult to measure, and no doubt would require in-depth interviews with key firms and institutions in the region. Also, we do not yet know how innovation processes and networks in the metropolitan region of Vienna compare to those found in other metropolitan regions of Europe, nor do we know their impact on the competitiveness of firms. The importance of these issues goes far beyond intellectual curiosity, as today economic competition affects firms in all areas far more than it did in the past, due to the accelerating pace of European integration and the pressures of global competition which are increasing the need for flexibility.

Acknowledgements: This contribution draws on work undertaken for an international project on 'The Regional Innovation Potential and Innovative Networks in Metropolitan Regions' of the German Research Association [DFG], under the responsibility of the Institute of Geography at the University of Hannover. That research is conducted through a partnership of four research institutions: the University of Hanover, the Royal Institute of Technology Stockholm, the Polytechnical University of Catalunia and the Austrian Academy of Sciences. The author of this contribution wishes to thank Javier Revilla-Diez (University of Hanover) for his comments at the various stages of the research. Vera Mayer, Walter Rohn (both of the Austrian Academy of Sciences) and Ingo Liefner (University of Hanover) have provided fundamental help in conducting the postal survey. Vera Mayer and Walter Rohn have assisted in computing the tables.

References

Andersson Å. (1985): Creativity and regional development, *Papers of the Regional Science Association* 56, 5-20

Chesnais F. (1988): Technical cooperation agreements between firms, *STI Review* 4, 51-120

Cohen W.M. and Levinthal D.A. (1989): Innovation and learning: The two faces of R&D, *Economic Journal* 99 (397), 569-596

DeBresson C. and Amesse F. (1991): Networks of innovators: A review and introduction to the issue, *Research Policy* 20 (5), 363-379

Dosi G. (1988): Sources, procedures, and microeconomic effects of innovation, *The Journal of Economic Literature* 26 (3), 1120-1126

Freeman C. (1991): Networks of innovators: A synthesis of research issues, *Research Policy* 20 (5), 499-514

Freeman C. (1986): The role of technical change in national economic development. In: Amin A. and Goddard J. (eds.) *Technological Change, Industrial Restructuring and Regional Development*, Allen & Unwin, London, pp. 100-114

Hall P. (1986): The theory and practice of innovation policy: An overview. In: Hall P. (ed.) *Technology Innovation and Economic Policy*, St. Martin's Press, New York, pp. 1-34

Johannisson B. (1991): University training for entrepreneurship: Swedish approaches, *Entrepreneurship and Regional Development* 3, 67-82

Kline S.J. and Rosenberg N. (1986): An overview of innovation. In: Landau R. and Rosenberg N. (eds.) *The Positive Sum Strategy*, National Academy Press, Washington, pp. 275-305

Lundvall B.-Å. (ed.) (1992): *National Systems of Innovation: Towards a Theory of Innovation and Interactive Learning*, Pinter, London

Lundvall B.-Å. (1988): Innovation as an interactive process: From user-producer interaction to the national system of innovations. In: Dosi G., Freeman C., Nelson R.R., Silverberg G. and Soete L. (eds.) *Technical Change and Economic Theory*, Pinter, London, pp. 349-369

Malecki E.J. (1997): *Technology & Economic Development. The Dynamics of Local, Regional and National Competitiveness*, Longman, Harlow, Essex

Malecki E.J. and Veldhoen M.E. (1993): Network activities, information and competitiveness in small firms, *Geografiska Annaler* 75B, 131-147

Mansfield E. (1968): *Industrial Research and Technological Changes*, W.W. Norton, New York

Mansfield E., Romeo A., Schwartz M., Teece D.J., Wagner S. and Brach P. (1982): *Technology Transfer, Productivity, and Economic Policy*, W.W. Norton, New York

Meyer-Krahmer F. (1985): Innovative behaviour and regional indigenous potential, *Regional Studies* 19 (6), 523-534

Myers M.B. and Rosenbloom R.S. (1996): Rethinking the role of industrial research. In: Rosenbloom R.S. and Spencer W.J. (eds.) *Engines of Innovation: US Industrial Research at the End of an Era*, Harvard Business School Press, Boston, pp. 209-228

Nelson R.R. and Winter S. (1982): *An Evolutionary Theory of Economic Change*, Belknap Press of Harvard University Press, Cambridge [MA]

OECD (1992): *Technology and Economy: The Key Relationships*, Organisation for Economic Co-operation and Development, Paris

Porter M.E. and Fuller M.B. (1986): Coalitions and global strategy. In: Porter M.E. (ed.) *Competition in Global Industries*, Harvard Business School Press, Boston, pp. 315-343

Powell W.W. (1990): Neither market nor hierarchy: Network forms of organization. In: Staw B.M. and Cummings L.L. (eds.) *Research in Organizational Behavior*, JAI Press, Greenwich [CT], pp. 295-335

Romer P. (1990): Endogenous technical progress, *Journal of Political Economy* 98, 71-103

Saviotti P.P. (1998): On the dynamics of appropriability, of tacit and of codified knowledge, *Research Policy* 26 (7/8), 843-856

Saviotti P.P. (1988): Information, entropy and variety in technoeconomic development, *Research Policy* 17, 89-103

Storper M. (1997): *The Regional World. Territorial Development in a Global World*, The Guilford Press, New York, London

Suarez-Villa L. (1989): *The Evolution of Regional Economies. Entrepreneurship and Macroeconomic Change*, Praeger, New York

Teece D.J. (1986): Profiting from technological innovation: Implications for integration, collaboration, licensing and public policy, *Research Policy* 15 (6), 285-305

Teece D.J. (1981): The market for know-how and the efficient international transfer of technology, *Annals of the American Academy of Political and Social Science* 458, 81-96

Tijssen R.J.W. (1998): Quantitative assessment of large heterogeneous R&D networks: The case of process engineering in the Netherlands, *Research Policy* 26 (7/8), 791-809

Zairi M. (1992): *Competitive Benchmarking: An Executive Guide. TQM Practitioner Series*, Technical Communications (Publishing) Ltd., Letchworth [UK]

7 Knowledge Interactions between Universities and Industry in Austria: Sectoral Patterns and Determinants

with *D. Schartinger, C. Rammer and J. Fröhlich*

The relationship between university and industry is a complex and heterogeneous phenomenon, and an important topic of the recent debate on innovation systems. This chapter explores the role of knowledge flows between universities and firms in the Austrian national innovation system and provides valuable insights into several dimensions of knowledge flows that are not typically explored in research on this topic. The patterns of interaction between 46 different fields of science and 49 economic sectors represent an important and interesting outcome of the analysis. Left censored Tobit models are used to evaluate the effect of sector specific and science field specific characteristics upon the probability of knowledge interaction, disaggregated by type of interaction.

1 Introduction

External links directed at the transfer of knowledge, are of crucial importance for the innovation performance of firms (see OECD 2000a; Dodgson 1994). The systems of innovation approach emphasises the importance of interactions among firms, public research institutions and technology policy for innovation success (Freeman 1987; Lundvall 1988, 1992; Nelson 1993; Nelson and Rosenberg 1993). Universities play three major roles within an innovation system (see Smith 1995): First, they undertake a general process of scientific research and thereby affect the technological frontier of industry over the long run. Second, they produce knowledge which is directly applicable to industrial production (prototypes, new processes etc.). Third, universities provide major inputs for industrial innovation processes in terms of human capital, either through the education of graduates, who become industry researchers or through personnel mobility from universities to firms.

This paper aims to contribute to the growing literature on university-industry linkages within an innovation system. In recent years, the focus of most of the studies on university-industry interactions has been based on detailed analysis of science-industry links in narrowly defined fields of research and technology (so-called 'high-tech' industries, see for example, Bania et al. 1993; Rees 1991; Ács et al. 1994; Meyer-Krahmer and Schmoch 1998), on the aggregate effect of university research on knowledge production in firms (Jaffe 1989; Anselin et al. 1997), or on certain types of knowledge interactions such as citations of university research in firm patents (Jaffe et al. 1993; Almeida and Kogut 1995), personnel mobility (Bania et al. 1992; Almeida and Kogut 1995; Hicks 2000), joint publications (Hicks et al. 1993, 1996; Hicks 2000) and spin-off formations of new firms by university members (OECD 2000b; Parker and Zilberman 1993; Kelly et al. 1992).

In this study, the focus is laid on a systematic analysis of university-industry interactions within a national innovation system, taking Austria as the empirical case. Within the framework of an interactive model of innovation (Kline 1985) we assume that universities contribute to industrial innovation not only by offering new kinds of technological development but via a variety of interactions. A main purpose is to explore the significance of the use of different channels of knowledge interactions between university and industry, and the sectoral differences which may occur. These sectoral differences in knowledge interactions are assumed to depend on size effects, proximity effects (i.e., technological proximity between university research and technology development in industry), and sector specific performance and structures. The analysis is based on a comprehensive data set which covers all fields of science and all sectors of economic activity and distinguishes different types of knowledge interactions.

The paper is organised as follows. The next section provides a basic account of different types of knowledge interaction between universities and firms. In Section 3 we move attention to variation in knowledge interactions at the level of fields of science and industry sectors. Section 4 briefly describes the data used to measure knowledge interaction. Section 5 makes a modest attempt to identify patterns of knowledge interaction between fields of science and sectors of economic activity, while Section 6 aims to explain the variations in the patterns identified. The concluding section summarises some of the major findings of the study.

2   Types of Knowledge Interactions between Universities and Industry

The notion knowledge interaction is used here to cover all types of direct and indirect, personal and non-personal interactions between firms and universities, directed at the exchange of knowledge within innovation processes. The channels used for transferring knowledge depend on the characteristics of knowledge, such as the degree of codification, the tacitness or the embeddedness in technological artefacts. The potential economic value of knowledge affects the way, knowledge is exchanged between actors (Saviotti 1998). Table 1 lists the types of knowledge interactions, which seem to be especially relevant in the case of university-industry interactions (see Schmoch et al. 2000). Knowledge interactions may be characterised by the degree of formalisation, the suitability for transferring tacit knowledge and the degree at which they are based on personal contacts (see Bonaccorsi and Piccaluga 1994):

- Direct personal interactions (i.e., face-to-face communication) build up social capital such as trust, a joint 'language' and a joint research culture. Social capital facilitates the transmission and exchange of information and knowledge because communication proceeds relatively smoothly (Boschma 1999). The establishment of social capital through personal interactions between university and firm members is likely to be facilitated by a common disciplinary background.

- Personal interactions are associated with the exchange of tacit knowledge through activities such as talking and listening or demonstrating and copying (Machlup 1980). What is considered as new knowledge, very often is the new combination of already existing pieces of knowledge. The combination occurs through personal interaction and communication processes between individuals. A special case of transfer of tacit knowledge between universities and firms is the mobility of graduates, i.e., the employment of graduates by firms. At the time of employment, graduates are equipped with a high amount of tacit knowledge, acquired at universities.

- Formalisation of the interaction is a different approach to ensure a sufficient level of trust and to reduce uncertainty. Formalisation of interaction has at least two functions: First, to commit human resources to objectives and views. Second, to avoid appropriability problems by choosing formal arrangements that correspond to needs of the partners involved (see Bonaccorsi and Piccaluga 1994).

- Furthermore, the length of interaction, the sequence of interaction and the resource involvement on both sides affect the type, volume and efficiency of knowledge exchange between university and industry.

The focus in the current study is on knowledge interactions that involve face-to-face-contacts and personal relations at least to some extent. These personal relations can take a wide variety of forms, reaching from occasional telephone and e-mail contacts over informal exchanges at forums and workshops to permanent structures such as university-industry research consortia and long-term collaborative research agreements. The focus on personal interactions rests on two arguments: *First*, personal interactions are a precondition for transferring tacit (not-codified) knowledge, which is regarded as particularly important for an effective knowledge exchange in innovation processes. A critical success factor in innovation is the capability of firms to find, select and absorb knowledge relevant for innovation (Cohen and Levinthal 1989, 1990). This capability is affected by the type of knowledge being exchanged (i.e., embodied or disembodied, codified or tacit), but in general eased if learning processes are stimulated by personal contacts (Foray and Lundvall 1996; David and Foray 1995). *Second*, personal interactions allow for the building up of trust between both partners. Trust again is a critical condition for co-operation, characterised by high uncertainty of results, the involvement of highly sensitive knowledge and a low exclusive appropriability of research results by one partner. It is worth noting that the focus is on knowledge interactions that have some degree of formalisation. This excludes informal meetings and casual meetings at conferences or various types of events where personal interactions and some kind of knowledge exchange may take place.

Table 1   Types of knowledge interactions between university and firms

Types of knowledge interaction	Formalisation of interaction	Transfer of tacit knowledge	Personal (face-to-face) contact
Employment of graduates by firms	+/–	+	–
Conferences or other events with firm and university participation	–	+/–	+
New firm formation by university members	+	+	+/–
Joint publications	–	+	+
Informal meetings, talks, communications	–	+	+
Joint supervision of PhDs and Master theses	+/–	+/–	+/–
Training of firm members	+/–	+/–	+
Mobility of researchers between universities and firms	+	+	+
Sabbatical periods for university members	+	+	+
Collaborative research, joint research programmes	+	+	+
Lectures at universities, held by firm members	+	+/–	+
Contract research and consulting	+	+/–	+
Use of university facilities by firms	+	–	–
Licensing of university patents by firms	+	–	–
Purchase of prototypes, developed at universities	+	–	–
Reading of publications, patents etc.	–	–	–

+ denotes interaction that typically involves formal agreements, transfer of tacit knowledge, personal contacts; +/– denotes interaction with a varying degree of formal agreements, transfer of tacit knowledge, personal contacts; – denotes interaction that typically does not involve formal agreements, transfer of tacit knowledge, and personal contacts.

3   Variation in Knowledge Interactions between University and Industry

We know from the results of past research that knowledge interactions between universities and industry are in fact complex because of different reasons. *First*, different types of knowledge interaction represent different strategies to keep up with two main prerequisites that condition industrial innovation processes: ensuring research efficiency (covering a wide range of skills, cost and risk reduction, exploitation of synergies, appropriation of returns from research) and getting access to scientific and technical opportunities (high quality researchers, research networks, instruments, new knowledge) (see Hicks at al. 1996; Katz and Martin 1997).

Second, intensity of knowledge flows may vary from type to type of knowledge interaction. Knowledge flows from university to industry are assumed to be strongest in the case of interactions that are based on close and recurring face-to-face contacts. This seems to be especially the case for joint research projects, joint publications, occupational mobility of academic researchers and spin-off formations of new firms. In the case of a joint supervision of Doctoral and Master theses, the face-to-face contact is maintained by a third party: the graduate or post-graduate student. With respect to contract research and consulting, this typically includes face-to-face-contacts at the beginning and the end of the contract, but does not have to involve face-to-face contacts in between.

Third, different types of knowledge interaction are associated with different types of personal relations. Joint publications and joint research projects are forms of collaboration where at least one university researcher and one industrial researcher co-operate. And in the case of spin-off formations of new enterprises, researchers, equipped with knowledge acquired at universities, attempt to commercialise this knowledge in setting up an own enterprise. *Finally*, types of knowledge interaction may differ with respect to the direction of the associated knowledge flows. In joint publications and joint research projects there are bi-directional flows of knowledge over a limited period of time (see Meyer-Krahmer and Schmoch 1998; Schmoch 1999). Spin-offs are associated with a bi-directional flow of knowledge if the university researcher remains at least partly assigned to the university but with a uni-directional if this is not the case.

In the current study, particular attention is paid to variation in the intensity of knowledge interactions between different sectors of economic activity and different fields of science. Although firms increasingly become 'multi-technological' (Pavitt 1998), not all research carried out at universities is equally relevant to all sectors of economic activity. Obviously a variety of factors do matter. There are firm and industry specific factors, affecting the type and amount of knowledge needed, how knowledge is conceived of, how the acquisition of knowledge is organised, who participates in the acquisition of knowledge etc. (Oinas 1999).

Different sectors of economic activity face different technological opportunities (Klevorick et al. 1995). They differ in the feasibility and sources of advance in their product and process relevant technologies. New scientific developments and results of university research constitute one source of new contributions to an economic sector's pool of technological opportunities. In their analysis, Klevorick et al. (1995) find that economic sectors, such as the production of drugs or food products, rely more heavily on university research than all others. Vice versa, certain fields of science, such as computer science or materials science, are highly relevant to a large number of sectors of industrial activity, whereas others, e.g. geology, bear high relevance only for a very limited number of industrial activities such as mining. It therefore seems interesting to analyse the pattern of linkages between university and industry not only at the micro level, but also at the levels of sectors of industrial activity and fields of university research.

There are at least two reasons why one may expect variations in sectoral patterns of knowledge interactions between universities and firms:

- *First*, two groups of sciences may be distinguished: 'pure sciences' and 'transfer sciences'. Pure sciences explore the boundaries of knowledge without concern for the practical implications of the findings. Transfer sciences which include the various branches of engineering share with the pure sciences a concern for predictive science, but otherwise they have rather different characteristics. Their activity is driven principally by the urge to solve problems arising from social and economic activities. A large part of their finding comes from industry. The play an essential role in providing an interface between the world of 'pure science' and the world of industry.

- *Second*, in different industries, innovation stems from very different sources within science and engineering. In some industries, firms make strong demands on technology provided by other industrial sectors. Others rely heavily on their own R&D efforts. R&D laboratories must also bring scientific and technological knowledge into the firm.

4   Measuring Knowledge Interactions

Knowledge creation and diffusion are intangible activities, which are difficult to measure. They are definitely not measurable in terms of what we normally think of as statistical variables (Grupp 1990). Due to the immaterial character of knowledge flows, empirical studies face considerable difficulties in their iden-tification and measurement. In response to these difficulties several analyses focus on those aspects of knowledge flows which are relatively easy to measure due to their explicit, codified character. Typical proxies to be used are citations of university publications in patents or publications by firms, licensing of university patents by firms, joint publications by firm and university members (see Hicks 2000; Jaffe 1989; Meyer-Krahmer and Schmoch 1998) or financial flows from industry to university in the case of contract research (see Siegel et al. 1999; Carlsson and Fridh 2000). A major shortcoming of these studies is the limited scope of knowledge flows covered. Various forms of personal contacts and the associated flows of tacit knowledge are left out of consideration. Knowledge interactions between industries and fields of science, which rest more on transfer mechanisms, strongly involving personal interactions, are covered to a lesser extent.

Another approach to measure knowledge interactions, is to ask researchers at industry and university about the types of interactions they use to exchange knowledge and about the significance of these types. By using a variety of indicators, different aspects of knowledge interactions and the corresponding flows of knowledge can be identified. A major shortcoming of this approach is, of course, the high degree of subjectivity. It is extremely difficult to compare individual perceptions and assessments on knowledge interactions.

In the current study we followed this approach to analyse knowledge interactions between university and industry on a broad, representative scale, i.e., knowledge interactions between all economic sectors and all fields of science within the Austrian innovation system. On the side of the industry, the business sector is disaggregated by 49 economic sectors, mainly corresponding to the NACE 2-digit classification. On the side of the universities, we distinguish 46 fields of science using a 2-digit classification which covers all academic disciplines, including humanities and social sciences (see Table 2). For each pair of industrial sector and field of science, the volume of knowledge interaction is measured for different types of knowledge interactions. The analysis rests on data from Austria for the second half of the 1990s.

Generally, knowledge interactions can be measured at the two sides of the interaction: the firm and the university. In this study, university departments (subdivisions of university institutes) are chosen as reporting units. The main source for empirical analysis is a survey of all departments at Austrian universities on various types of knowledge interactions with firms. The heads of departments were asked to provide information on the following nine types of personal knowledge interactions (the reference period being 1995 to 1998):

- the number of collaborative research projects, carried out together with firms,
- the number of scientific publications, jointly written with firm members,
- the number of researchers who changed to firms for the purpose of R&D activities (either temporarily or permanently),
- the number of technology-oriented firms, founded by members of the department,
- the number of Doctoral and Master theses, jointly supervised with members of a firm or carried out at firms,
- the number of lectures by firm members,
- the number of training courses for firm members, offered by the department,
- the number of research assistants, financed by firms.

Table 2   List of sectors of economic activity and fields of science

Sectors of economic activity	Industry group	Fields of science	Faculty
01 Agriculture	Resource int.	11 Mathematics, informatics	Natural sciences
02 Forestry	Resource int.	12 Chemistry	Natural sciences
05 Fishing	Resource int.	13 Physics, mechanics, astronomy	Natural sciences
10 Mining of coal	Resource int.	14 Biology, botanics, zoology	Natural sciences
11 Extract. of oil and gas	Resource int.	15 Geology, mineralogy	Natural sciences
12/13 Mining of metal ores	Resource int.	16 Meteorology, climatology	Natural sciences
14 Other Mining and quarrying	Resource int.	17 Hydrology, hydrography	Natural sciences
15/16 Manuf. of food, beverages, tabacco	Labour int.	18 Geography	Natural sciences
17 Manuf. of textiles	Labour int.	19 Other, interdisciplinary natural sciences	Natural sciences
18 Manuf. of wearing apparel	Labour int.	21 Mining, metallurgy	Techn. sciences
19 Manuf. of leather products	Labour int.	22 Engineering	Techn. sciences
20 Manuf. of wood, wood products	Resource int.	23 Construction techniques	Techn. sciences
21 Manuf. of paper, paper products	Resource int.	24 Architecture	Techn. sciences
22 Printing and publishing	Labour int.	25 Electrical engineering	Techn. sciences
23 Manuf. of coke, refined petroleum	Resource int.	26 Technical chemistry	Techn. sciences
24 Manuf. of chemicals	Hum.cap. int.	27 Geodesy	Techn. sciences
25 Manuf. of rubber and plastic products	Labour int.	28 Traffic and transport science	Techn. sciences
26 Manuf. of mineral products	Resource int.	29 Other, interdisciplin. technical sciences	Techn. sciences
27 Manuf. of basic metals	Resource int.	31 Anatomy, pathology	Medicine
28 Manuf. of fabricated metal products	Labour int.	32 Medical chemistry, physics, physiology	Medicine
29 Manuf. of machinery	Hum.cap. int.	33 Pharmacy, pharmacology, toxicology	Medicine
30 Manuf. of computers, office machinery	Hum.cap. int.	34 Hygiene, medical microbiology	Medicine
31 Manuf. of electronical machinery	Hum.cap. int.	35 Clinical medicine	Medicine
32 Manuf. of electronics	Hum.cap. int.	36 Surgery, anaesthesiology	Medicine
33 Manuf. of medical, optical, precis. instr.	Hum.cap. int.	37 Psychiatry, neurology	Medicine
34 Manuf. of motor vehicles	Hum.cap. int.	38 Forensic medicine	Medicine
35 Manuf. of other transport equipment	Hum.cap. int.	39 Other, interdisciplinary human medicine	Medicine
36 Manuf. of furniture and others	Labour int.	41 Agriculture	Agric. sciences
37 Recycling	Resource int.	42 Horticulture	Agric. sciences
40 Production and supply of energy	Resource int.	43 Forestry	Agric. sciences
41 Production and supply of water	Resource int.	44 Animal production	Agric. sciences
45 Construction	Labour int.	45 Veterinary medicine	Agric. sciences
50 Trade and repair of motor vehicles	Distribut. serv.	49 Other, interdisciplinary agric. sciences	Agric. sciences
51 Wholesale trade	Distribut. serv.	51 Political science	Social sciences
52 Retail trade	Distribut. serv.	52 Jurisprudence	Social sciences
55 Hotels and restaurants	Distribut. serv.	53 Economics	Social sciences
60 Road and rail transport services	Distribut. serv.	54 Sociology	Social sciences
61 Water transport services	Distribut. serv.	55 Psychology	Social sciences
62 Air transport services	Distribut. serv.	56 Spatial Planning	Social sciences
63 Travel agencies etc.	Distribut. serv.	57 Applied statistics, social statistics	Social sciences
64 Post and telecommunication services	Distribut. serv.	58 Educational science	Social sciences
65 Financial intermediation	Prod.rel. serv.	59 Other, interdisciplinary social sciences	Social sciences
66 Insurance	Prod.rel. serv.	61 Philosophy	Humanities
67 Auxil. activities to financial intermed.	Prod.rel. serv.	64 Theology	Humanities
70/71 Real estate, rental services	Prod.rel. serv.	65 Historical sciences	Humanities
72 Software and related activities	Prod.rel. serv.	66 Linguistics, literature	Humanities
73 Research and Development	Prod.rel. serv.	67 Other, philological-cultural studies	Humanities
74 Business services	Prod.rel. serv.	68 Fine arts	Humanities
90 Waste water and refuse services	Distribut. serv.	69 Other, interdisciplinary humanities	Humanities

For each type of interaction, respondents had to report the respective number of interactions for each out of 49 economic sectors (supported by a list and definition of sectors). The term 'firm' was defined as private business enterprise from the agricultural, manufacturing and service sectors. Each university department (the smallest academic unit at Austrian universities, generally with one full professor and some assistants) was assigned to one of 46 fields of science, allowing for the construction of knowledge interaction matrices between economic sectors and fields of science for each type of interaction.

Information was gathered by means of a standardised questionnaire and a postal survey. The reporting units were university departments at all 12 Austrian universities (including three technical universities, one economic university, one university for agricultural sciences and one university for veterinary medicine, but excluding schools of art and technical colleges). For larger departments with a sub-department organisation, these sub-departments were chosen as reporting units in order to improve validity and reliability of responses. A complete list of all departments and sub-departments was provided by the Federal Ministry for Science and Research. The survey was conducted in spring 1999. 421 questionnaires were returned, which is equal to a response rate of 37.2 percent in terms of the total number of departments and 48.3 percent in terms of the total number of researchers at university departments. Each field of science shows a response rate of at least 20 percent (see Table 3 for more details).

A second source of information covers formal research projects between industry and university. Two databases are available containing information on research projects carried out by university departments of all the 12 Austrian universities. One database, called FODOK, comprises 8,916 research projects for the period of time from 1983 to 1996. Another database, called AURIS, contains 8,145 research projects from 1996 to 1999. The two databases differ in their structure and in some aspects of the information gathered and that fact makes it impossible to simply add the number of research projects. Research projects, carried out together with business partners, can be identified via the entry 'project partners'. University departments either had to report the name of the partner-company or the economic sector, the co-operating partner was belonging to. A matching procedure of names of project-partners with a company database allowed for identification of partners from the business sector and an assignment to economic sectors (for technical details see Schartinger et al. 2000). The information from these databases can be taken as a proxy for university-industry relations in the field of contract research and consulting.

This approach, used for collecting information on knowledge interactions between university and industry, has some weaknesses, of course. *First*, the counting data for each type of knowledge interaction may represent very different volumes of interaction, depending e.g. on the size of the research projects, the length of co-operation and the frequency of contacts. *Second*, each type of interaction indicates very different types of associated knowledge flows, differing in quantity, quality and relevance for innovation activities. *Third*, only looking at respondents from one side of the interaction may produce biased results. A survey of Austrian innovative firms on their interaction behaviour with universities shows that at least the relevance of the various interaction channels does not differ significantly from the pattern reported by university departments (see Schartinger et al. 2001). Although the absolute numbers of knowledge interactions observed in our survey may be inaccurate, structural characteristics of interaction patterns, such as the sectoral decomposition and the specialisation on different channels of knowledge interaction, can be analysed, however, and should produce valid results.

Table 3   Response rates of the 1999 survey of university departments, differentiated by fields of science

Fields of science	Surveyed departments in % of total departments	Academic personnel in surveyed departments in % of total academic personnel
Natural sciences		
11 Mathematics, informatics	30.8	39.7
12 Chemistry	56.3	53.8
13 Physics, mechanics, astronomy	20.4	23.2
14 Biology, botanics, zoology	55.2	70.2
15 Geology, mineralogy	46.7	52.9
16 Meteorology, climatology	66.7	86.7
17 Hydrology, hydrography	33.3	51.8
18 Geography	66.7	54.9
19 Other, interdisciplinary natural sciences	25.0	34.6
Technical sciences		
21 Mining, metallurgy	41.5	59.0
22 Engineering	31.3	35.0
23 Construction techniques	35.5	36.5
24 Architecture	25.0	30.2
25 Electrical engineering	37.0	54.6
26 Technical chemistry	24.0	18.8
27 Geodesy	22.2	33.3
28 Traffic and transport science	35.4	45.0
29 Other, interdisciplinary technical sciences	50.0	26.4
Human medicine		
31 Anatomy, pathology	28.6	32.9
32 Medical chemistry, physics, physiology	39.1	49.1
33 Pharmacy, pharmacology, toxicology	46.2	54.1
34 Hygiene, medical microbiology	19.1	16.7
35 Clinical medicine	71.4	66.9
36 Surgery, anaesthesiology	45.7	82.9
37 Psychiatry, neurology	25.0	32.6
38 Forensic medicine	40.0	43.8
39 Other, interdisciplinary human medicine	25.0	37.8
Agriculture and forestry, veterinary medicine		
41 Agriculture	27.5	34.6
42 Horticulture	20.0	22.5
43 Forestry	36.9	44.9
44 Animal production	25.0	18.5
45 Veterinary medicine	25.0	20.7
49 Other, interdisciplinary agricultural sciences	0.0	0.0
Social sciences		
51 Political science	40.0	60.0
52 Jurisprudence	30.6	39.1
53 Economics	35.2	46.9
54 Sociology	56.7	61.2
55 Psychology	25.0	32.4
56 Spatial planning	40.0	57.7
57 Applied statistics, social statistics	24.3	26.2
58 Educational science	50.0	74.6
59 Other, interdisciplinary social sciences	25.0	43.7
Humanities		
61 Philosophy	44.4	41.3
64 Theology	41.3	39.7
65 Historical sciences	51.9	51.2
66 Linguistics, literature	40.4	54.7
67 Other philological-cultural studies	100.0	100.0
68 Fine arts	35.3	37.5
69 Other, interdisciplinary humanities	9.1	8.3
Total	37.2	48.3

5   Sectoral Patterns of Knowledge Interactions

The sectoral pattern of knowledge interactions and variation in this pattern between distinct types of interaction is described by matrices of interaction ($^t m_{rs}$) where m represents the share of knowledge interactions I in total number of knowledge interactions of type t; $s = 1, ..., S$ denotes the economic sector, $r = 1, ..., R$ the field of science and $t = 1, ..., T$ the type of knowledge interaction. The matrix ($^t m_{rs}$) shows where knowledge interactions between fields of science and sectors of economic activity do cluster and it shows, where no interaction takes place between fields of science and sectors of economic activity. The elements of the matrix are defined by

$$^t m_{rs} = \frac{^t I_{rs}}{\sum\limits_{r=1}^{R}\sum\limits_{s=1}^{S} {}^t I_{rs}}. \tag{1}$$

One may assume that more interactions will take place between larger fields of science and larger economic sectors than between smaller ones, simply because of the size. Assuming that this size effect depends on the research potential in each field and sector, respectively, one can compute an 'expected market share' p_{rs} for $r = 1, ..., R$ and $s = 1, ..., S$. Research potential is measured by the number of researchers (full-time equivalent) P so that

$$p_{rs} = \frac{(P_r\, P_s)^{\frac{1}{2}}}{\sum\limits_{r=1}^{R}\sum\limits_{s=1}^{S}(P_r\, P_s)^{\frac{1}{2}}}. \tag{2}$$

Normalising the interaction share $^t m_{rs}$ by the expected market share $^t p_{rs}$ produces a normalised interaction share $^t n_{rs}$, which is a measure for a size adjusted intensity of interaction. Values above one indicate that interactions between a science field r and an economic sector s are more intense than one would expect because of their size:

$$^t n_{rs} = \frac{^t m_{rs}}{p_{rs}}. \tag{3}$$

In order to give an indication of the overall sectoral pattern of knowledge interactions between fields of science and economic sectors, we produce a matrix of interaction aggregated over $t = 1, ..., T$. Matrices for specific types of knowledge interactions are weighted by a specific constant for these interactions $^t w$ ($0 \le {}^t w \le 1$, $\Sigma_t\, {}^t w = 1$) and are summed up for all types of knowledge interaction t. The weight w represents the relative significance (considering both, quantity and quality) of type t of knowledge interaction as a transfer channel in the innovation process as perceived by firms. Data for these weights stem from a survey of innovative Austrian firms on their co-operation behaviour in innovation. Firms had to assess the relative significance of each type of interaction for knowledge exchange with universities in the context of innovation activities (see Schibany et al. 1999; Schartinger et al. 2001). The type specific weights are given in Table 5. The size adjusted interactive share n_{rs} may be calculated as

$$n_{rs} = \frac{\sum_{t=1}^{T} {}^t w \, {}^t m_{rs}}{p_{rs}}. \tag{4}$$

Table 4   Characteristics of faculties in Austria in 1995[a]

| Faculty | Personnel | Share in total number of | | | |
		Public funding	Research projects[b]	Research projects with industry[c]	Knowledge interaction with industry[d]
Natural sciences	23	29	37	25	28
Technical sciences	11	11	20	32	37
Medicine	31	32	17	19	10
Agriculture, forestry; veterinary medicine	3	4	4	3	4
Social sciences	19	13	14	16	19
Humanities	12	11	9	4	2
Total	100	100	100	100	100

[a] Figures in %; Source: Austrian Federal Ministry for Science and Research, survey by the authors, own calculations; [b] research projects funded, by general university funds, by the FWF Austrian Science Fund, other public and private funds; [c] measured by the projects, financed by a public fund that supports industrial innovation projects and projects, financed by 'others'; as all public financing sources are covered in separate categories, this category is assumed to comprehend projects, financed by private sources; [d] measured by weighted aggregated interaction shares n_{rs} (see Equation (4)) considering nine types of knowledge interactions (joint research projects, joint publications, contract research, personnel mobility by university researchers, financing of research assistants, joint supervision of theses, training for firm members, lectures by firm members, start-ups by university researchers)

Table 5   Relative importance of type of interaction for knowledge exchange between industry and university as revealed by firms

Type of knowledge interaction	Normalised share of innovative firms stating that they use the channel of interaction for knowledge exchange with universities
Collaborative research	0.15
Joint scientific publications	0.10
Contract research and consulting	0.22
Mobility of researchers from universities to firms	0.06
Financing of university research assistants by firms	0.14
Joint supervision of Doctoral and Master theses	0.14
Training of firm members at universities	0.07
Lectures at universities held by firm members	0.07
Technology oriented new firm formation by university researchers	0.05
Total	1.00

Source: Schartinger et al. (2000, 2001), Schibany et al. (1999), own calculations

Figure 1 shows the aggregated interaction matrix for these size adjusted interaction shares n_{rs}. The totals for each field of science (n_r) and each economic sector (n_s) is a measure for the intensity of knowledge interactions to industry and university, respectively, with respect to the size of a field or sector. For comparison, both the share in total interactions (m_r and m_s) and the share in total R&D personnel

$$p_r = \frac{P_r}{\sum_r P_r} \quad \text{and} \quad p_s = \frac{P_s}{\sum_s P_s} \tag{5}$$

is given in the last column and last row, respectively, in Figure 1. Matrix cells are shaded according to their shares. Economic sectors and fields of science are ordered by their scientific orientation and business orientation, respectively. Sectors of economic activity were listed by the decreasing share of academic employees in total employment (as a proxy for science orientation), while fields of science by the decreasing share of graduates that were employed in the private business sector (the remaining share is employed in the state sector, i.e., public administration and public services including health and education). All matrix cells, which show a result greater than 0.4, are shaded. The darker the cells the more intense are the corresponding knowledge interactions between an economic sector and a field of science.

One may expect that interactions will occur more often between highly scientific and business orientated sectors and fields, respectively. Thus, high intensities of interaction should cluster in the left upper area of the matrix of interaction. The descriptive analysis only partially confirms this. Knowledge interactions between fields of science and economic sectors are disperse and do not follow obvious and simple patterns. Looking at the overall intensities of interaction by fields of science and sectors of economic activities, results correspond to common knowledge on university-industry interaction. Among the ten fields of science with the highest intensity of interaction, there are mainly technical disciplines (mining & metallurgy, construction techniques, other technical disciplines) and natural sciences (physics, mathematics & informatics, chemistry) but also economics and other social sciences. About 70 percent of all knowledge interactions with the industry are performed by these fields of science. Humanities and medical sciences show substantially lower intensities of inter-action.

On the side of the industry, the list of sectors with highest intensity of interaction only partially corresponds with common rankings of knowledge intensive sectors. Within the group of sectors with high R&D ratios, only the sectors research and development, chemical industry, manufacturing of instruments and motor vehicles are among the ten sectors with the highest interaction intensities. Electrical industry and electronics and manufacturing of machinery – which cover about 45 percent of all R&D expenditures in the Austrian private business sector – show a lower level of knowledge interaction than one would expect with respect to their R&D potential. In the case of electronics industry, this may be explained by the high share of foreign capital in this sector and thus strong external knowledge links, i.e., knowledge interaction with partners outside the Austrian innovation system.

High intensities of interaction can be observed in more resource oriented manufacturing sectors such as energy production, manufacturing of basic metals, manufacturing of paper, construction and agriculture. The comparatively strong scientific orientation of these traditional manufacturing sectors corresponds with an international technological specialisation of Austrian manufacturing in these sectors revealed by patent statistics (see Fischer et al. 1994). In the service sector, intensities of interaction follow a pattern that one would expect, with high intensities in producer related services, banking, insurance and computer services. A special case is the printing and publishing sector where the development of new publication series and journals is viewed as product innovations introduced in co-operation with universities.

The matrix of interactions also allows to analyse the spread of knowledge flows, i.e., whether fields of science interact with a broad range of economic sectors or not. Variations in the range of economic sectors that fields of science interact with have two implications: First, the variations reveal the extent to which innovation activities in economic sectors rely on certain fields of science. Second, as fields of science are funded by the public, differences in sectoral linkages of fields of science cause that changes in the allocation of public funding for universities have sectorally different effects.

Among the fields of science which interact intensively with a broad range of sectors (measured by the variance of normalised interaction shares n_{rs} when controlling for interaction intensity of a field of science n_r) are economics, physics, mathematics and informatics, electrical engineering, other technical disciplines and – to a lesser extent – chemistry, biology and engineering. On the other side, fields of science with strong linkages to only a few sectors of economic activities are mining and metallurgy, chemical engineering, mechanical engineering, pharmaceutics, clinical medicine, transportation sciences, botany and zoology.

From the viewpoint of economic sectors, one may distinguish sectors, which rely on a broader spectrum of scientific knowledge in their innovation processes from sectors, which show a high concentration in terms of scientific knowledge. Among the former group one finds chemical industry, manufacturing of machinery, instruments, energy production, computer services, R&D and producer related services. These sectors interact with a broad range of fields of science. This indicates their reliance on a broader spectrum of scientific knowledge. The opposite is true for agriculture, mining, food production, paper production, manufacturing of basic metals, manufacturing of motor vehicles, and insurance. Knowledge interaction by these sectors is concentrated on a few fields of science only.

There are particularly strong university-industry linkages (if controlled by size) between some fields of science in technical sciences and some manufacturing sectors. Such concentrated interaction patterns appear between chemical engineering and manufacturing of paper, botany and agriculture; forestry science

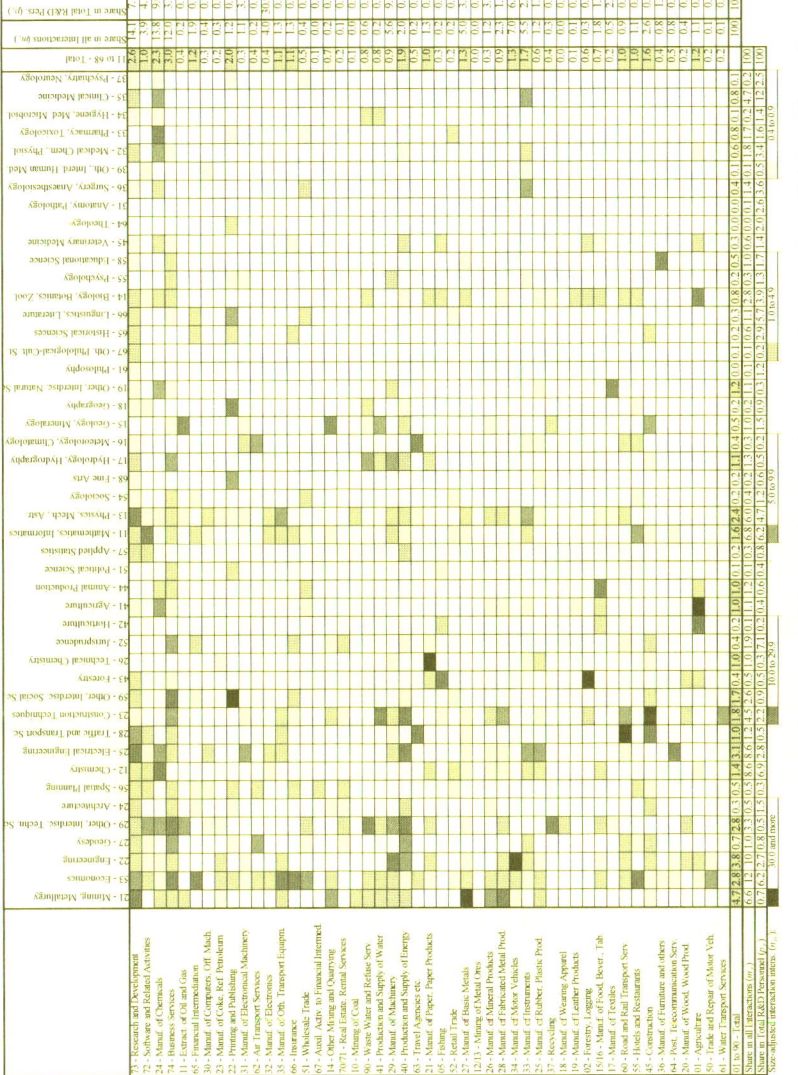

Figure 1 Knowledge interactions between fields of science and sectors of economic activities: Size adjusted interactions shares, aggregated for nine types of interactions

and forestry/wood industry; mining and geology; metallurgy and manufacturing of basic metals. These are examples for a well established division of labour between public research, where public research provides infrastructure and expertise directly oriented at the demand for knowledge and technology of certain industries in their innovation activities.

In order to describe variations in the type of interaction we define an index for the deviation of average, size adjusted intensities of interaction of different types of interactions t from the aggregated intensity of interaction for each field of science ($^t q_r$) and each sector of economic activity ($^t q_s$). To ease interpretation, we focus on four types of interactions t: contract research ($t = I$), joint research ($t = II$; covering both, joint R&D projects and joint publications), personnel mobility ($t = III$; including temporary and permanent mobility of university researchers to firms, financing of university research assistants by firms, and joint supervision of Doctoral and Master theses), and co-operation in education ($t = IV$; including training courses at universities for firm members as well as lecturers by firm members at universities). The deviation indices, $^t q_r$ and $^t q_s$, are transformed in a way that they vary between −1 and +1:

$$
^t q_r = \tanh\left[\ln\left(\frac{\sum_{s=1}^{S} {}^t n_{rs}}{\sum_{s=1}^{S} n_{rs}}\right)\right] \quad \text{and} \quad {}^t q_s = \tanh\left[\ln\left(\frac{\sum_{r=1}^{R} {}^t n_{rs}}{\sum_{r=1}^{R} n_{rs}}\right)\right]. \tag{6}
$$

Figure 2 shows deviations of interaction intensities by type of interaction for twelve fields of science and twelve industrial sectors with the highest share in the total number of interaction, respectively (m_r, m_s). Figure 2 demonstrates that the use of different types of knowledge interaction varies significantly between fields and sectors. Joint research is a way of knowledge interaction used predominantly by natural and technical sciences (engineering, chemistry, physics) but of minor relevance in the social sciences. On the one side, manufacturing sectors such as chemical industry, instruments, basic metals and vehicles use this type of interaction more intensively. On the other side, many economic sectors, which engage in joint research above average, do so in contract research activities below average (e.g. chemical industry and vehicle industry) and vice versa. The same is true on the level of fields of science (e.g. mining/metallurgy and chemistry).

For the chemical and vehicle industries, personnel mobility is also an intensively used type of knowledge interaction, with correspondingly high intensities of interaction in personnel mobility in the fields of science of chemistry and mechanical engineering. Personnel mobility is also an important mechanism for the transfer of knowledge in economics. Training courses for firms and lectures by firm members are, however, the most important type of interaction used by economics and other social sciences. On the part of the economic sectors, the service industry acquires knowledge from universities via training and education to a higher extent than other sectors.

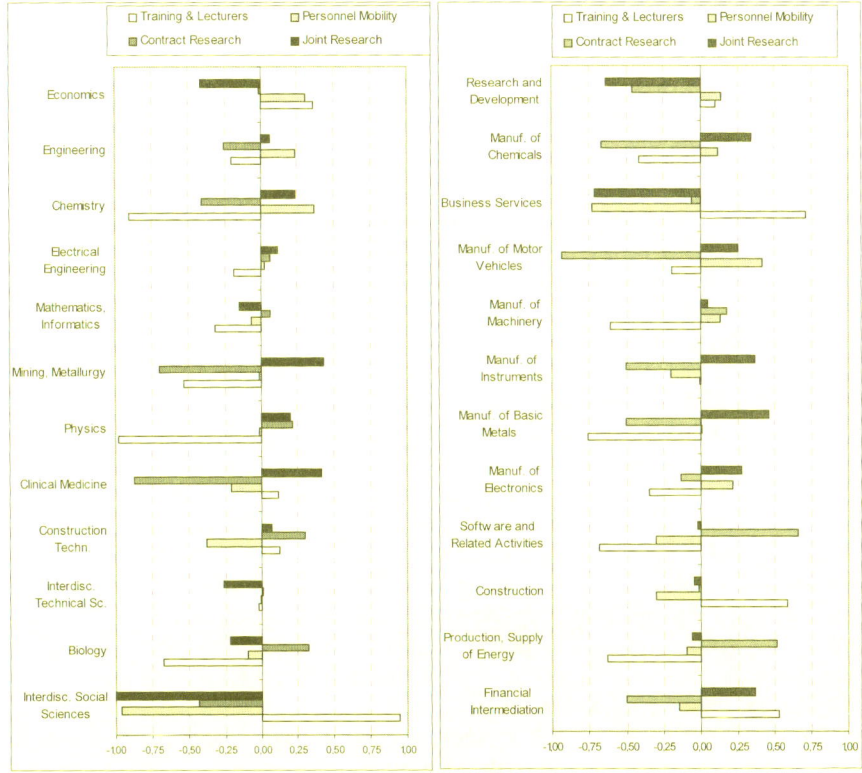

Figure 2 Variations in the use of different types of knowledge interaction by fields of science and sectors of economic activity

6 Determinants of Knowledge Interactions

In this section we attempt to identify determinants which explain the variation in the patterns of knowledge interaction between fields of science and sectors of economic activity, described in the previous section. From a modelling perspective, the relative significance of knowledge interactions $^t m_{rs}$ between a field of science r and a sector of economic activity s for a certain type of interaction t may be modelled as a function of the size of each field (that is, P_r) and sector (that is, P_s), respectively, c_r and c_s representing structural characteristics of field of science r and sector of economic activity s, respectively, which affect the capability and propensity to engage in knowledge interactions, and d_{rs} representing distance (in terms of spatial and 'knowledge' proximity) between a field of science r and a sector s of economic activity. In order to analyse the variation in the effects of size, proximity and structural characteristics on sectoral knowledge interaction patterns for certain types of interaction, the same set of variables is used for each type t of knowledge interaction:

$$^t m_{rs} = {}^t f(P_r, P_s, c_r, c_s, d_{rs}). \tag{7}$$

The size of fields and sectors is measured in terms of R&D personnel. The number of researchers is an indicator for the potential of personal knowledge interactions and is supposed to positively affect the level of interaction.

Proximity between a field of science and a sector of economic activity is linked to two different concepts of proximity for knowledge interaction: spatial proximity and knowledge proximity. Several studies have shown that geographical location of universities is a major factor, which influences the behaviour of interaction (Mansfield and Lee 1996; Jaffe et al. 1993; Mansfield 1995, 1997). Although communication technologies significantly improved communication over long distances, face-to-face contacts are still of central importance in knowledge interactions. Costs of face-to-face contacts, however, heavily depend on the physical distance between the locations of the contact partners (see Batten et al. 1989). 'Spatial distance' between a field of science and a sector of economic activity is measured as the average traveling distance (shortest road distance in km) between each university department that belongs to a certain science field r and a firm that belongs to a certain economic sector s, departments and firms being weighted by their share in total employment in r and s, respectively.

Although firms become increasingly 'multi-technological' (Pavitt 1998), not all the research carried out at universities is equally interesting for all economic sectors and vice versa. Therefore the relevance of university research to solve a technological problem is highly dependent on the knowledge proximity of fields of knowledge and sectors of applications. In the case of engineering or information science, for example, one may assume strong ties to industry and therefore a high propensity to exchange knowledge while for humanities this propensity will be significantly lower.

In order to measure knowledge proximity, we use the share of graduates educated in science field r and working in economic sector s as a proxy. This share includes all employed graduates (i.e., also those graduated in the past). Thus, this share is an indicator for longer-term relations in human capital. Although this is a crude proxy for knowledge proximity, it covers major aspects of this concept. As knowledge is mainly incorporated in human capital, i.e., employees, personnel mobility between university and industry is likely to be the main transfer mechanism for knowledge generated in field r and economically used in sector s. The long-term pattern of such knowledge relations should therefore represent differences in knowledge proximity by fields of science and economic sectors. There are, however, factors such as idiosyncratic job decisions and employment of graduates outside the fields of study which may disturb this pattern. Nevertheless, we assume that there is no systematic variation of these factors by fields of science and industrial sectors.

In order to explain patterns of knowledge interaction between university and firms, special attention is paid to those structural characteristics of fields of science and economic sectors which represent (a) their attractiveness as a source of knowledge and a partner for knowledge exchange and (b) their capacities to engage in knowledge interactions and to absorb the knowledge provided by and jointly produced with a partner from university or industry. On the side of the university, five variables are used to reflect these aspects:

- *Size of a field of science*: The size of a field of science (measured in terms of average academic staff per department) may strongly affect the resources available for R&D projects aside from day-to-day tasks (such as the education of students and organisational tasks). It may be assumed that on the one hand a predominance of very small departments is likely to offer favourable prerequisites for interactions due to flexibility and specialisation on focussed research fields. On the other hand, fields of science dominated by large departments may provide more extensive technical infrastructure, which may ease collaboration. A medium sized size structure is viewed as being less flexible than a small one while providing less infrastructure than a large one. Thus, it is expected that the curve of size effects takes a U-shaped form.

- *Experience in external R&D collaboration*: If there is a certain level of experience in external, industry oriented knowledge interactions in a certain field of science, institutional and individual barriers to knowledge interactions are likely to be less important than in the case of fields of science with little experiences so far. Moreover, previous knowledge interactions by university departments will expand the contact network relevant for knowledge spillover to industry and thus increase the probability of future interactions.

- *Reputation*: A likely precondition for a field of science at a university to be used as interaction partner by industry is the acknowledgement of expertise of its members. Out of the three main activities that universities in Austria have to devote resources to – research, teaching and administration – only research activities result in the development of expertise and scientific excellence, which renders knowledge exchange profitable for their partners. In order to reduce risk and costs of knowledge interactions, industry will tend to look for university departments offering a higher quality of research output. A commonly used indicator for the quality of research output is publications in international, refereed journals. Thus, research reputation is measured by the number of such publications per researcher. Mansfield and Lee (1996) have shown that faculties with a distinctively high reputation have a much higher propensity to acquire R&D support from firms far away than do faculties with a low reputation.

- *Employment dynamics*: Increasing employment in a certain field of science can not solely be linked to an increased demand in labour, as it is the case in industry. It also reflects a political desire to support certain fields of science.

These may seem especially beneficial for the growth of social welfare in creating scientific inputs and qualified labour supply for economic sectors that are perceived to be of technological and economic importance by political actors. Under the assumptions that political actors foresee those fields of science correctly that are of 'strategic' importance for developing certain economic sectors, the growth of a field of science at a university may be associated with increasing links to industry.

- *Intensity of marketing activities*: It seems likely that every way of informing a broader public of research activities and results is conducive in order to establish knowledge interactions. In this context, the presence of university research results in the public domain in form of presentations and statements, publications in print media with high pervasion or TV and radio broadcastings is considered advantageous for knowledge interactions between the university and the business sector to take place.

On the side of the industry, four structural characteristics that attempt to cover knowledge absorption capacities and the demand for strong scientific linkages are taken into account:

- *Size structure of the economic sector*: It seems to be a robust empirical pattern that R&D increases according to the size of the firm and therefore enables firms to 'plug in' to external sources of scientific and technological expertise (Cohen 1996). This plugging in only becomes possible if the firm is equipped with a stock of knowledge in a particular domain that conditions its ability to evaluate and exploit extra firm sources of knowledge, i.e., its absorption capacity. As it is easier for larger firms to maintain in-house R&D facilities, this is clearly an argument for larger firms, establishing more external linkages. The size structure of a sector of economic activity is measured by the employment share of large firms (500 and more employees) and the employment share of small firms (less than 100 employees).

- *R&D orientation of the economic sector*: In order to transfer knowledge from universities to firms, firms ought to have some capacity to absorb knowledge. This absorption capacity (Cohen and Levinthal 1989, 1990) is highly dependent on learning experiences in the past which are likely to increase with a higher research orientation of the firm. The concept of absorption capacity implies that in order to have access to a piece of knowledge, developed elsewhere, it is necessary to have experience in R&D on something similar (Saviotti 1998). Thus, R&D may be viewed as serving a dual, but strongly interrelated role: first, developing new products and production processes and second, enhancing the learning capacity (Fischer 2001). Critical indicators for the R&D orientation of a firm sector are its R&D ratio and its share of R&D personnel.

- *Export orientation of the economic sector*: The exposition of sectors to international competition, measured in terms of export ratios, is likely to force these sectors to engage more actively in innovation activities. It may be argued that this comparably increased pressure to innovate motivates firms to generally establish more external linkages, as firms are known to rarely innovate in isolation (OECD 1998). Interactions with universities are also likely to rise more in these economic sectors rather than in sectors that mainly produce for national markets.

- *Employment dynamics of the economic sector*: High dynamics in economic sectors are assumed to reflect the growing labour demand of sectors with young, fast-moving and radically new technologies. These sectors are particularly dependent on technological innovations and scientific progress, and therefore are inclined to engage in knowledge interactions with universities more than others.

In order to control for variations in framework conditions in fields of science and groups of industry we include fields of science specific and economic sector specific dummies into the interaction models. On the one side, dummies are used for natural sciences, technical sciences, economics, social sciences (excl. economics), medicine and agricultural sciences (with reference to humanities). On the other side, dummies for labour intensive manufacturing, human capital intensive manufacturing, resource intensive manufacturing and producer oriented services (with reference to distribution services) shall reflect technological and market characteristics which affect interaction behaviour and which are not seized by the variables discussed above. The assignment of science fields and industrial sectors to the faculties and industry groups, respectively, is shown in Table 2.

Information on fields of science is derived from biannual reports by Austrian university departments, covering all departments from all universities. The data is collected and provided by the Austrian Federal Ministry for Science and Research and include, among others, the number of academic staff (differentiated by qualification), the number of publications in national and international journals, the number of research projects carried out (by source of financing), the number of graduate and postgraduate students, whose Master or Doctoral theses are supervised by the department. For our study we used average data for the reporting periods 1990-1995 (assuming that there is a time lag between the performance of a field of science and its perception by industry). Data on economic sectors is taken from the 1995 census on economic activity and from other data sources, provided by the Austrian Statistical Office. For details on data sources and descriptive statistics on dependent and independent variables see Table 6. A correlation matrix of model variables is given in Table 7.

Table 6   Dependent and independent variables, used in the knowledge interaction models

	Mean	Standard deviation	Minimum	Maximum	Source[a]
Knowledge interactions I – contract research: share of interactions between r and s in total number of contract research projects 1990-99 ($^{I}m_{rs}$)	0.044	0.204	0	4.85	Auris, Fodok
Knowledge interactions II – joint research: share of interactions between r and s in total number of joint R&D projects and joint publications 1995-98 ($^{II}m_{rs}$)	0.044	0.357	0	9.87	Survey
Knowledge interactions III – mobility: share of interactions between r and s in total number of researcher mobility, research assistant financing and joint supervision of Doctoral and Master theses 1995-98 ($^{III}m_{rs}$)	0.044	0.312	0	8.13	Survey
Knowledge interactions IV – training: share of interactions between r and s in total number of lectures by firm members and training courses for firm members 1995-98 ($^{IV}m_{rs}$)	0.044	0.403	0	11.24	Survey
Knowledge interactions V – aggregate index: share of weighted interactions of all types between r and s in total number of weighted interactions of all types ($^{V}m_{rs}$)	0.044	0.251	0	6.13	Survey, Auris, Fodok
Size of r [ln of number of researchers in 1994-1995 (full professors, associated professors, research assistants)] ($\ln P_r$)	4.719	1.034	3.04	6.91	BMWV
Size of s [ln of number of research personnel in 1995 (at full time equivalents – FTE)] ($\ln P_s$)	4.426	1.764	1.61	8.65	ÖSTAT(3)
'Knowledge proximity' (share of graduates in r working in s from all employed graduates in 1991) ($^{K}d_{rs}$)	0.044	0.237	0	4.78	ÖSTAT(1)
Average spatial distance (average road distance in km from each research location of r, weighted by distance to each firm of s within Austria in 1991) ($^{S}d_{rs}$)	195.1	38.6	17.4	285.9	ÖSTAT(2)
Size structure of r [I, average department size in 1994-1995 (number of researchers at FTE per department)] ($^{I}ST_r$)	10.17	5.18	3.4	35.9	BMWV
Size structure of r (II, exponent of average department size in 1994-1995) ($^{II}ST_r$)	130.3	189.6	11.5	1290.9	BMWV
Experience in contract research at r (number of research contracts per researcher, average for 1990-95) (RE_r)	0.378	0.386	0.04	1.94	BMWV
Research reputation of r (number of publications in international journals per researcher, average for 1990-95) (RR_r)	0.979	0.818	0.002	3.27	BMWV
Employment dynamics in r (number of researchers in 1995 per number of researcher in 1990) (ED_r)	1.153	0.206	0.77	2.33	BMWV
'Marketing activities' by r (number of PR-activities per researcher, average for 1990-1995) (MA_r)	1.980	1.312	0.35	6.08	BMWV
Size structure of s (I, employment share of firms > 500 employees in 1995) ($^{I}ST_s$)	0.301	0.275	0	0.93	ÖSTAT(3)
Size structure of s (II, employment share of firms < 250 employees in 1995) ($^{II}ST_s$)	0.455	0.286	0	0.99	ÖSTAT(3)
Research orientation of s (R&D expenditures as a share of value, added in 1995) (RO_s)	0.036	0.088	0	0.57	ÖSTAT(3)
Export orientation of s (exports as a share of turnover in 1995) (EO_s)	0.246	0.261	0	0.94	ÖSTAT(3) (4)
Employment dynamics in s (number of employees in 1995 per number of employees in 1991) (ED_s)	1.020	0.185	0.54	1.50	ÖSTAT(2) (3)

[a] Auris: Auris-Database (1996-1999), Fodok: Fodok-Database (1983-96) on research projects by Austrian University Departments; Survey: survey by the authors in 1999 at Austrian University Departments on knowledge interactions with industry; BMWV: Federal Ministry for Science and Research, 'Arbeitsberichte der Institutsvorstände' (1990-91, 1992-93, 1994-95); ÖSTAT(1): Austrian Statistical Office, 'Volkszählung 1991'; ÖSTAT(2): Austrian Statistical Office, 'Arbeitsstättenzählung 1991'; ÖSTAT(3): Austrian Statistical Office, 'Bereichszählung 1995'; ÖSTAT(4): Austrian Statistical Office, 'Außenhandelsstatistik 1995'

Table 7   Correlation matrix of interaction model variables[a]

	$^{I}m_{r_s}$	$^{II}m_{r_s}$	$^{III}m_{r_s}$	$^{IV}m_{r_s}$	$^{V}m_{r_s}$	$\ln P_r$	$\ln P_s$	$^{k}d_o$	$^{s}d_o$	$^{I}ST_r$	$^{II}ST_r$	RE_r	RR_s	ED_r	MA_s	$^{I}ST_s$	$^{II}ST_s$	RO_s	EO_s	ED_s
$^{I}m_{r_s}$	1.00																			
$^{II}m_{r_s}$	0.41	1.00																		
$^{III}m_{r_s}$	0.49	0.86	1.00																	
$^{IV}m_{r_s}$	0.25	0.40	0.41	1.00																
$^{V}m_{r_s}$	0.62	0.90	0.03	0.61	1.00															
$\ln P_r$	0.14	0.09	0.12	0.08	0.13	1.00														
$\ln P_s$	0.23	0.15	0.15	0.10	0.19	0.17	1.00													
$^{k}d_o$	0.22	0.17	0.18	0.37	0.29	0.05	0.09	1.00												
$^{s}d_o$	−0.13	−0.06	−0.07	−0.03	−0.08	0.05	0.05	−0.03	1.00											
$^{I}ST_r$	0.03	0.06	0.05	0.01	0.05	0.53	–	−0.03	0.03	1.00										
$^{II}ST_r$	0.01	0.06	0.04	0.01	0.04	0.46	–	−0.03	0.04	0.94	1.00									
RE_r	0.06	0.06	0.07	0.02	0.07	−0.24	–	0.00	−0.25	−0.13	−0.14	1.00								
RR_s	0.01	0.04	0.03	−0.04	0.02	0.32	–	−0.09	0.11	0.43	0.36	−0.15	1.00							
ED_r	−0.01	0.01	0.01	−0.01	0.01	−0.19	–	−0.03	−0.16	0.11	0.05	0.09	−0.14	1.00						
MA_s	−0.08	−0.06	−0.07	−0.01	−0.07	−0.32	–	−0.03	0.20	−0.42	−0.30	−0.21	−0.17	−0.24	1.00					
$^{I}ST_s$	−0.03	0.03	0.02	−0.02	0.00	–	0.20	−0.02	−0.21	–	–	–	–	–	–	1.00				
$^{II}ST_s$	−0.02	−0.07	−0.06	0.01	−0.04	–	−0.31	0.04	0.19	–	–	–	–	–	–	−0.86	1.00			
RO_s	0.38	0.13	0.13	0.04	0.18	–	0.48	0.02	−0.20	–	–	–	–	–	–	−0.02	−0.11	1.00		
EO_s	0.06	0.12	0.10	0.02	0.10	–	0.65	−0.04	0.32	–	–	–	–	–	–	0.11	−0.31	0.24	1.00	
ED_s	0.07	0.02	0.03	0.05	0.05	–	0.16	0.08	−0.20	–	–	–	–	–	–	−0.09	0.18	0.16	−0.19	1.00

[a] excluding dummy variables; variables' abbreviations are explained in Table 4. Source: Austrian Federal Ministry for Science and Research, survey by the authors, own calculation

Dependent variables (interaction shares by type, measured in percent) may either have the value zero or a metric value between zero and one hundred, the largest observed value is 11.24%. The share of observations with positive interactions ranges from 7% to 30%, depending on the type of interaction. For this kind of data, censored regression models are an appropriate modelling approach. We use left-censored Tobit models to estimate a model of type (7). Empirical estimations were carried out with Stata 5.0. Table 8 reports the estimation results.

The size of a field of science and a sector of economic activity (measured in terms of research personnel), as well as the knowledge proximity between a field of science and an economic sector (measured by the share of graduates, educated in a certain science field and working in a certain economic sector), explain a significant part of the variance in interaction shares between industry sectors and fields of science. Coefficients for structural characteristics of fields of science are not statistically significant, except for experience in contract research, which shows the expected positive sign. Fields of science, which belong to natural sciences, technical sciences, agricultural sciences and economics show significantly higher propensities to interact with industry than medicine, social sciences (excl. economics) and humanities.

On the side of the industry, however, knowledge interactions do not vary among economic sectors, which are broadly defined by factor intensities. High R&D intensities and high employment dynamics (i.e., expanding markets and/or high competitiveness of industries) are the major factors, which affect the interaction share of economic sectors. Furthermore, the size structure of the industry influences knowledge interaction activities with universities. The propensity of interaction is lower in sectors with a high share of large firms.

The estimation results for each of the four types of knowledge interaction distinguished show significant variations in the effects of independent variables and allow for some qualification of the mechanism behind these effects (see Table 8). Only in the case of contract research, spatial distance does affect interactions negatively. Contract research (including consulting) especially involves sectors and fields of science with low interaction activities, minor or no use of other transfer channels and presumably irregular knowledge interactions with industry or university, respectively. For these fields and sectors, contract research may be viewed as an interaction type with low entry costs, requiring comparably low absorption and transfer capacities at the own side. Potential interaction partners are searched for at the nearby university or within the regional economy. If universities and firms are separated by long distance, the probability to find a potential partner and establish a contact will be lower. For the three other types of knowledge interaction, which all necessarily demand direct personal relationships and/or a higher extent of trust, distance seems to be no barrier for interactions, especially when taking into account that Austria's extension is rather small.

Experience in contract research increases the probability of knowledge interaction between a field of science and industry especially in the case of contract research and personnel mobility. Both types of interaction demand a significant amount of administrative arrangements and legal agreements. University

Table 8 Parameter estimates of left-censored Tobit-models for knowledge interactions between fields of science and sectors of economic activity[a]

	Aggregated interactions		Joint research		Contract research		Personnel mobility		Training & lectures	
Size of r	0.160	(8.26)**	0.555	(4.18)**	0.152	(8.73)**	0.551	(7.11)**	0.507	(4.30)**
Size of s	0.110	(8.41)**	0.344	(4.12)**	0.103	(9.10)**	0.272	(5.82)**	0.306	(3.97)**
Knowledge proximity	0.371	(8.64)**	1.047	(5.52)**	0.225	(6.14)**	0.455	(4.06)**	1.177	(6.47)**
Average spatial distance	-0.0007	(-1.38)	0.0002	(0.07)	-0.0008	(-1.82)*	0.0014	(0.79)	-0.0005	(-0.18)
Size structure of r (I, average department size)	-0.008	(-0.73)	-0.0009]	(-0.12)	-0.0007	(-0.07)	0.017	(0.42)	-0.031	(-0.44)
Size structure of r (II, exponent of average department size)	0.0002	(0.87)	0.0006	(0.38)	0.0004	(0.17)	-0.0005	(-0.58)	0.0006	(0.37)
Experience in contract research in r	0.122	(2.61)**	0.480	(1.59)	0.083	(2.05)**	0.540	(3.05)**	0.422	(1.52)
Research reputation of r	0.001	(0.05)	0.412	(2.31)**	-0.012	(-0.48)	-0.012	(-0.12)	-0.167	(-0.95)
Employment dynamics in r	0.025	(0.28)	0.001	(0.00)	-0.015	(-0.20)	0.353	(1.17)	-0.370	(-0.57)
Marketing activities in r	0.017	(1.06)	-0.007	(-0.06)	0.017	(1.26)	-0.013	(-0.22)	-0.021	(-0.21)
Size structure of s: Share of large firms	-0.226	(-1.88)*	-1.468	(-2.01)**	-0.267	(-2.58)**	-0.690	(-1.62)	-0.521	(-0.71)
Size structure of s: Share of small firms	-0.040	(-0.34)	-1.413	(-1.92)*	-0.030	(-0.29)	-0.624	(-1.47)	-0.333	(-0.45)
Research intensity of s	0.406	(2.76)**	1.329	(1.53)	0.902	(7.18)**	0.312	(0.64)	0.340	(0.43)
Export orientation of s	-0.094	(-1.06)	-0.382	(-0.70)	-0.146	(-1.86)*	-0.362	(-1.17)	-1.024	(-1.79)*
Employment dynamics in s	0.230	(2.53)**	0.208	(0.36)	0.123	(1.56)	0.534	(1.61)	1.091	(1.97)**
Faculties										
Natural sciences	0.295	(5.34)**	1.110	(2.75)**	0.287	(5.83)**	0.728	(3.54)**	0.763	(2.32)**
Technical sciences	0.348	(5.92)**	1.697	(3.95)**	0.343	(6.72)**	0.733	(3.33)**	1.023	(2.94)**
Medicine	0.048	(0.62)	-0.363	(-0.65)	0.076	(1.09)	-0.171	(-0.57)	0.093	(0.19)
Agricul. sciences, forestry, veterinary medicine	0.216	(3.07)**	1.702	(3.49)**	0.203	(3.26)**	0.854	(3.18)**	0.604	(1.35)
Economics	0.225	(3.87)**	1.164	(3.03)**	0.184	(3.47)**	0.862	(4.51)**	1.055	(3.47)**
Social sciences (excl. Economics)	0.052	(0.90)	0.484	(1.11)	-0.011	(-0.20)	0.180	(0.82)	0.650	(1.99)**
Sectors of economic activities										
Labour intensive manufacturing	0.035	(0.65)	-0.295	(-0.85)	0.057	(1.19)	-0.250	(-1.27)	0.011	(0.03)
Human capital intensive manufacturing	0.058	(0.95)	0.006	(0.01)	0.0026	(0.05)	-0.022	(-0.10)	0.375	(1.05)
Resource intensive manufacturing	0.064	(1.38)	-0.080	(-0.27)	0.087	(2.07)**	0.017	(0.10)	0.066	(0.24)
Producer oriented services	0.034	(0.65)	-0.564	(-1.55)	0.055	(1.16)	-0.048	(-0.26)	0.292	(0.96)
Constant	-1.94	(-7.73)**	-7.72	(-4.67)**	-1.67	(-7.64)**	-6.99	(-6.98)**	-7.22	(-4.47)**
Log likelihood	-846.4		-516.3		-682.2		-645.0		-573.0	
Pseudo R^2	0.29		0.21		0.36		0.23		0.18	
Total number of observations	2.254		2.254		2.254		2.254		2.254	
Number of uncensored (positive) observations	665		144		577		232		159	

[a] *t*-values in parentheses. Source: survey 1999, own calculations; * statistically significant at the 0.1 level; ** statistically significant at the 0.05 level

researchers who have not made any experience in these fields so far, may perceive them as barriers and avoid such investments for setting up a co-operation with the industry. In the case of joint research and publication, scientific quality of research, carried out in a field of science, positively affects the probability of interacting with industry. This result confirms that the industry attempts to reduce uncertainty in the outcome of joint research efforts with the university by selecting partners with high reputation.

On the part of the industry, the size structure of economic sectors shows significant effects on the propensity to engage in knowledge interactions with universities in joint research and contract research. Results of estimations suggest that sectors with high shares of medium sized enterprises (which are – in the Austrian context – firms with more than 100 but less than 500 employees) have ceteris paribus higher interaction activities in joint research with universities than other sectors. In the case of contract research, sectors with low shares of large firms show ceteris paribus higher intensities of interaction. The negative effect of the share of large firms may be due to a loose international orientation of larger firms within a small and open economy such as in Austria. As these firms, and the sectors in which they are concentrated, have a high international orientation and high shares of foreign capital, knowledge acquisition and co-operation in innovation rest on foreign sources to a higher extent[16]. This effect of a comparably loose anchoring in the national innovation system of the internationally oriented sectors in a small and open economy is also revealed by the negative coefficient for the export ratio of a sector, which is statistically significant in the case of contract research and training.

The research intensity of a sector, i.e., the share of R&D expenditures in total turnover, is a major determining factor for a sector's interaction activities with universities in terms of contract research. Knowledge exchange by the means of personnel mobility and training is, however, not affected by research intensity. These types of interaction demand less absorption capacities on the side of the industry and are therefore used by sectors with low R&D orientation to a similar extent as by research intensive sectors. Sectors, which are growing more rapidly in terms of employment use universities as a source for the transfer of knowledge via training more frequently than sectors with low growth rates.

A surprising result is the low influence of a sector's research intensity on the extent of collaborative research and joint publication with universities. This may again be caused by a particular sectoral structure of the Austrian economy with the two most research intensive sectors (chemistry and drugs, electronics and telecommunication equipment), being of high international orientation and showing – with respect to their R&D potential and intensity – comparably low knowledge interaction activities within the Austrian innovation system.

---

[16] This has been revealed by the Austrian results from the Community Innovation Surveys in 1996 (CIS-II), which report a share of 50 to 60 per cent for small and medium sized firms (both for manufacturing and the service sector) co-operating in innovation with partners from abroad, while this share is 75 to 80 per cent for large firms (defined as those firms with more than 250 employees).

7   Summary and Conclusions

The objective of this study was to analyse certain aspects of knowledge interaction between universities and industry, concentrating on nine types of knowledge interaction, which involve direct personal relations. Special emphasis was laid on sectoral differences, distinguishing 46 fields of science and 49 sectors of economic activities. Empirical data was gathered by a survey of all Austrian universities, thus covering knowledge interactions for the whole Austrian innovation system on a representative basis.

Although the results show a number of peculiarities which are associated with the particular situation in Austria as a small and open economy with a predominantly small and medium sized firm structure, there are some results which seem to be of more general importance.

Apparently, universities and the industry use a variety of channels in order to transfer knowledge. The channels vary in the intensity of personal relations, in the types of knowledge transferred and in the direction of the knowledge flow. From the viewpoint of industry, the use of different channels represents varying strategies to ensure research efficiency, allows access to different types of scientific and technological knowledge and reflects differences in demand for knowledge in different stages of the innovation process. Sectors of economic activity and fields of science engage in different types of interactions. While technical sciences and R&D intensive manufacturing industries tend to use direct research co-operation more intensively, service industries and social sciences rest more on personnel mobility and training related interactions. Joint research and contract research seem to be used for opposite needs as fields of science and economic sectors which are strongly engaged in one of these types of knowledge interaction tend to engage in the other types far below average.

Some economic sectors interact with partners from a broad range of fields of science whereas others show a high concentration on partners in particular fields of science. This indicates that in some economic sectors a wide range of technologies and, consequently, a high degree of interdisciplinarity is necessary in order to develop innovative and competitive products. Contrarily, concentrated interaction patterns reflect a long-term, well established division of labour between public research, providing a research infrastructure and expertise, directly oriented at the demand for knowledge and technology of certain industries in their innovation activities.

Patterns of knowledge interaction between fields of science and economic sectors are strongly affected by the size of both, fields and sectors, and their knowledge proximity. On the side of the university, even when controlling for size and knowledge proximity, faculties show significantly different interaction activities with industries. Natural sciences, technical sciences, agricultural sciences and economics have higher intensities of interaction than those in medicine, social sciences (excl. economics) and humanities. Among the structural characteristics of science fields, the level of experience in contract research and scientific quality of research positively affect knowledge interaction with industry.

On the industry side, a high share of medium sized firms in a sector, a high R&D intensity and high employment dynamics exert a positive influence on the propensity to engage in knowledge interactions with universities. On the other hand, a high export orientation has a negative effect indicating the increased use of knowledge from abroad in the case of internationally oriented sectors. However, determinants vary for different types of knowledge interactions.

This study confirms that knowledge interaction between industry and university shows a complex pattern. *First*, interactions are not restricted to a few industries and science fields. Rather, a large number of scientific disciplines and almost all sectors of economic activities exchange knowledge in the course of industrial innovation. *Second*, R&D resources in industry and orientation of science fields towards industry application do not prejudice the level of knowledge interaction. Both, some traditional manufacturing and service sectors, and some basic research oriented science fields engage significantly in innovation related knowledge interaction with universities and industry, respectively. *Third*, industry and university use a large variety of channels for knowledge interaction. A restriction of the analysis of university-industry relations to only a few types of channels may produce misleading results as there are significant differences in the orientation on certain types of interaction by industrial sectors and fields of science. Looking only at one channel, evidenced, for example, by citations of university publications in firm patents or by financial flows for contract research projects, fades out and leads to distorted pictures of university-industry relations.

With respect to research policy, the variety of sectoral patterns and interaction types should be taken into account in the evaluation of university-industry relations and in debates on public promotion for fostering knowledge and technology transfer between industry and public research. Technology transfer to industry receives increasing attention in evaluating university performance. In many cases, indicators concentrate on only a few types of interactions such as research contracts and patent applications. Our results suggest to widen the set of indicators and to consider knowledge interactions such as training, personnel mobility, start-ups as well as other forms of personal contacts more seriously. In the area of public promotion programmes for university-industry relations there is a strong focus on supporting co-operation in research projects. While there is no doubt that knowledge exchange in direct research collaboration is one of the most effective transfer channels, other types of interaction should not be underestimated in their role in innovation processes. In the design of research and technology policy, barriers to university-industry interactions in other areas than direct research co-operation should be taken into account and addressed in effective ways as well.

Acknowledgements: The authors would like to thank two anonymous referees for their valuable comments. Furthermore, we are grateful to Andreas Schibany and Thomas Rödiger-Schluga for their comments on an earlier version of this paper. This study was funded by the FWF Austrian Science Fund.

References

Ács Z., Fitzroy F. and Smith I., (1994): High technology employment and university R&D spillovers: Evidence from US cities. Paper presented at the 41st Annual North American Meetings of the Regional Science Association International, Niagara Falls

Almeida P. and Kogut B. (1995): The geographic localization of ideas and the mobility of patent holders, Paper presented at the Conference on "Small and Medium-Sized Enterprises and the Global Economy", organised by CIBER, University of Maryland, October 20, 1995

Anselin L., Varga A. and Ács Z. (1997): Local geographic spillovers between university research and high technology innovations, *Journal of Urban Economics* 42 (3), 422-448

Bania N., Eberts R. and Fogarty M. (1993): Universities and the startup of new companies: Can we generalise from Route 128 and Silicon Valley? *The Review of Economics and Statistics* 75 (4), 761-766

Bania N., Calkins L. and Dalenberg R. (1992): The effects of regional science and technology policy on the geographic distribution of industrial R&D laboratories, *Journal of Regional Science* 32 (2), 209-228

Batten D.F., Kobayashi K. and Andersson Å.E. (1989): Knowledge, nodes and networks: An analytical perspective. In: Andersson Å.E., Batten D.F. and Karlsson C. (eds.) *Knowledge and Industrial Organization*, Springer, Berlin, Heidelberg, New York, pp. 31-46

Bonaccorsi A. and Piccaluga A. (1994): A theoretical framework for the evaluation of university-industry relationships, *R&D Management* 24 (2), 229-247

Boschma R. (1999): Culture of trust and regional development: An empirical analysis of the Third Italy, Paper presented at the 39th European Congress of the Regional Science Association 1999, 23 - 27 August 1999, Dublin, Ireland

Bozeman B. (2000): Technology transfer and public policy: A review of research and theory, *Research Policy* 29 (4/5), 627-655

Carlsson B. and Fridh A.C. (2000): Technology transfer in United States universities: A survey and statistical analysis, Department of Economics, Weatherhead School of Management, Cleveland

Cohen W.M. (1996): Empirical studies of innovative activity. In: Stoneman, P. (ed.) *Handbook of the Economics of Innovation and Technological Change*, Blackwell, Oxford [UK], Cambridge [MA], pp. 182-264

Cohen W.M. and Levinthal D.A. (1990): Absorptive capacity: A new perspective on learning and innovation, *Administrative Science Quarterly* 35, 128-152

Cohen W.M. and Levinthal D.A. (1989): Innovation and learning: The two faces of R&D, *Economic Journal* 99 (397), 569-596

David P.A. and Foray D. (1995): Accessing and expanding the science and technology knowledge base, *STI-Review* 16, 13-68

Dodgson M. (1994): Technological collaboration and innovation. In: Dodgson M. and Rothwell R. (eds.) *The Handbook of Industrial Innovation*, Edward Elgar, Chelterham [UK], Brookfield [VT], pp. 285-292

Fischer M.M. (2001): Innovation, knowledge creation and systems of innovation, *The Annals of Regional Science* 35 (2), 199-216

Fischer M.M., Fröhlich J. and Gassler H. (1994): An exploration into the determinants of patent activities. Some empirical evidence for Austria, *Regional Studies* 28 (1), 1-12

Foray D. (1997): Generation and distribution of technological knowledge: Incentives, norms and institutions. In: Edquist C. (ed.) *Systems of Innovation: Technologies, Institutions and Organizations*, Pinter, London, pp. 65-85

Foray D. (1994): Production and distribution of knowledge in the new Systems of innovation: The role of intellectual property rights, *STI-Review* 14, 119-152

Foray D. and Lundvall B.-Å. (1996): The knowledge-based economy: From the economics of knowledge to the learning economy. In: OECD (ed.) *Employment and Growth in the Knowledge-based Economy*, Organisation for Economic Co-operation and Development, Paris, pp. 11-32

Freeman C. (1987): *Japan: A New National System of Innovation?* Pinter, London

Grupp H. (1990): Technometrics as a missing link in science and technology indicators. In: Sigurdson J. (ed.) *Measuring the Dynamics of Technological Change*, Pinter, London, pp. 57-76

Hicks D. (2000): Using innovation indicators for assessing the efficiency of industry-science relationships. Paper presented at the Joint German-OECD Conference "Benchmarking Industry-Science Relations", Berlin, 16-17 October 2000

Hicks D., Isard P. and Martin B. (1996): A morphology of Japanese and European corporate research networks, *Research Policy* 25 (3), 359-378

Hicks D., Isard P. and Martin B. (1993): University-industry alliances as revealed by joint publications, Mimeo, SPRU

Jaffe A.B. (1989): Real effects of academic research, *American Economic Review* 79 (5), 957-970

Jaffe A.M., Trajtenberg M. and Henderson R. (1993): Geographic localisation of knowledge spillovers as evidenced by patent citations, *Quarterly Journal of Economics* 108 (3), 577-598

Joly P.B. and Mangematin V. (1996): Profile of public laboratories, industrial partnerships and organisation of R&D: The dynamics of industrial relationships in a large research organisation, *Research Policy* 25 (6), 901-922

Katz J.S. and Martin B.R. (1997): What is research collaboration? *Research Policy* 26 (1), 1-18

Kelly K.-J., Weber J., Friend S., Atchinson G., DeGeorge W. and Holstein W. (1992): Hot spots. America's new growth regions are blossoming despite the slump, *Business Week* October 29, 80-88

Klevorick A.K., Levin R.C., Nelson R.R. and Winter S.G. (1995): On the sources and significance of interindustry differences in technological opportunities, *Research Policy* 24 (2), 185-205

Kline S.J. (1985): Innovation is not a linear process, *Research Management* 28, 36-45

Lundvall B.-Å. (ed.) (1992): *National Systems of Innovation. Towards a Theory of Innovation and Interactive Learning*, Pinter, London

Lundvall B.-Å. (1988): Innovation as an interactive process: From user-producer interaction to the national system of innovation. In: Dosi G., Freeman C., Nelson R.R., Silverberg G. and Soete L. (eds.) *Technical Change and Economic Theory*, Pinter, London, pp. 349-369

Machlup F. (1980): *Knowledge: Its Creation, Distribution and Economic Significance. Knowledge and Knowledge Production, Volume 1*, Princeton University, Princeton

MacPherson A. (1997): The role of producer-service outsourcing in the innovation performance of New York State manufacturing firms, *Annals of the Association of American Geographers* 87 (1), 52-71

Mansfield E. (1997): Links between academic research and industrial innovations. In: David P. and Steinmueller E. (eds.) *A Production Tension: University-Industry*

Collaboration in the Era of Knowledge-Based Economic Development, Stanford University Press, Palo Alto

Mansfield E. (1995): Academic research underlying industrial innovations: Sources, characteristics, and financing, *The Review of Economics and Statistics* 77 (1), 55-65

Mansfield E. and Lee J.-Y. (1996): The modern university: Contributor to industrial innovation and recipient of industrial R&D support, *Research Policy* 25 (7), 1047-1058

Meyer-Krahmer F. and Schmoch U. (1998): Science-based technologies: University-industry interactions in four fields, *Research Policy* 27 (8), 835-851

Nelson R.R. (ed.) (1993): *National Systems of Innovation: A Comparative Study*, Oxford University Press, Oxford [UK]

Nelson R.R. and Rosenberg N. (1993): Technical innovation and national systems. In: Nelson R.R. (ed.) *National Systems of Innovation: A Comparative Study*, Oxford University Press, Oxford [UK], pp. 3-21

OECD (2000a): Science, technology and industry outlook 2000: Innovation networks, DSTI/STP/TIP(2000)5, Organisation for Economic Co-operation and Development, Paris

OECD (2000b): Analytical report on high tech spin-offs, DSTI/STP/TIP(2000)7, Organisation for Economic Co-operation and Development, Paris

OECD (1998): National innovation systems: Analytical findings, DSTI/STP/TIP(98)6, Organisation for Economic Co-operation and Development, Paris

Oinas P. (1999): The difference that space makes in organisational learning. Paper presented at the 39th European Congress of the Regional Science Association 1999, 23-27 August 1999, Dublin, Ireland

Oliver N. and Blakeborough M. (1998): The multi-firm new product development process. In: Grieve-Smith J. and Michie J. (eds.) *Innovation, Cooperation and Growth*, Oxford University Press, Oxford [UK], pp. 151-160

Parker D. and Zilberman D. (1993): University technology transfers: Impacts on local and U.S. economies, *Contemporary Policy Issues* 11, 87-99

Pavitt K. (1998): The social shaping of the national science base, *Research Policy* 27 (8), 793-805

Rees G. (1991): New information technologies and vocational education and training in the new information technologies in the European Community, Background Report [Commission of the European Communities], Cardiff

Rothwell R. (1991): External networking and innovation in small and medium-sized manufacturing firms in Europe, *Technovation* 11, 93-112.

Saviotti P.P. (1998): On the dynamics of appropriability, of tacit and codified knowledge, *Research Policy* 26 (7/8), 843 -856

Schartinger D., Schibany A. and Gassler H. (2001): Interactive relations between university and industry: Empirical evidence for Austria, *Journal of Technology Transfer* 26 (3), 255-268

Schartinger D., Rammer C., Fischer M.M. and Fröhlich J. (2000): The role of space in research co-operation between university and firms, Discussion Paper, Department of Systems Research, Austrian Research Centers Seibersdorf, Austria

Schibany A., Jörg L. and Polt W. (1999): Towards realistic expectations. The science system as a contributor to industrial innovation. Study by the Working Association TIP in Commission of the Austrian Federal Ministries of Economic Affairs and Science and Transport, Department of Systems Research, Austrian Research Centers Seibersdorf, Austria

Schmoch U. (1999): Interaction of universities and industrial enterprises in Germany and the United States. A comparison, *Industry and Innovation* 6 (1), 51-68

Schmoch U., Licht G. and Reinhard M. (2000): *Wissens- und Technologietransfer in Deutschland*, IRB Verlag, Stuttgart

Siegel D., Waldman D. and Link A. (1999): Assessing the impact of organizational practices on the productivity of university technology transfer offices: An exploratory study, NBER Working Paper 7256, Cambridge [MA]

Smith K. (1995): Interactions in knowledge systems: Foundations, policy implications and empirical methods, *STI-Review* 16, 69-102

Part III

Knowledge Creation, Diffusion and Spillovers

8 Innovation, Knowledge Creation and Systems of Innovation

This chapter critically reviews the systems of innovation approach as a flexible and useful conceptual framework for spatial innovation analysis. It presents an effort to develop some missing links and to decrease the conceptual noise often present in the discussions on national innovation systems. The paper specifies elements and relations that seem to be essential to the conceptual core of the framework and argues that there is no a priori reason to emphasise the national over the subnational (regional) scale as an appropriate mode for analysis, irrespective of time and place. Localised input-output relations between the actors of the system, knowledge spillovers and their untraded interdependencies lie at the centre of the argument.

1 Introduction

At the turn of the century the world economy is undergoing a process of profound restructuring. Three developments have set the stage for this process. The first is a technological revolution, centering initially on telecommunications and microprocessors, but now extending to biological science that has created new industries and changed the methods that many established industries employ for production and distribution. The second has been a managerial revolution, initially associated with the diffusion of Japanese techniques for quality control, team production and supplier relations, but now extending to many innovative forms of production employed around the world including increasing individualisation and diversification of working relationships. The third development, reinforcing the first two, is the considerable empowerment of capital accompanied by a decline of the influence of labour unions. Under these conditions nation states and regions come increasingly under stress to sustain competitiveness and, thus, economic welfare.

Today, it is widely recognised that technological change is the primary engine for economic development. Innovation – the heart of technological change – is essentially the innovation process that depends upon the accumulation and development of relevant knowledge of a wide variety. Certainly individual firms play a crucial role in the development of specific innovations but the process that nurtures and disseminates technological change involves a complex web of interactions among a range of firms, other organisations and institutions.

The systems of innovation approach has recently received considerable attention as a promising conceptual framework for advancing our understanding of the innovation process in the economy. The approach contrasts with previous attempts such as the traditional OECD approach to technological change and innovation that focused on the R&D system in the narrower sense, primarily by analysing resource inputs and outputs of the system. A too narrow focus on R&D overlooks the importance of other types of innovative efforts in the business sectors and, thus, the innovative performance of low-tech sectors in the economy.

The main objective of this paper is to provide greater understanding of the approach as a flexible and useful conceptual framework for innovation analysis. It presents an effort to develop some missing elements and to decrease the conceptual noise often present in the discussions. The subsequent section of the paper examines the characteristics of the innovation process: its nature, sources and some of the factors that shape its development. It aims to introduce the reader to concepts that are central to understanding the systems approach and to emphasise the role of knowledge creation and dissemination.

A system of innovation may be thought of as a set of actors such as firms, other organisations, and institutions that interact in the generation, diffusion and use of new – and economically useful – knowledge in the production process. Institutions may be viewed as sets of common habits, routines, established practices, rules or laws that regulate the relations and interactions between individuals within as well as between and outside the organisations. Section 3 specifies elements and relations that are essential to the conceptual core of the systems of innovation approach. Territorially based systems build on spatial proximity – as either regional (subnational), national or global systems of innovation. Current research practice is focusing almost exclusively on the national scale in systems of innovation research (see, for example, Lundvall 1992, Nelson 1993, OECD 1994, Edquist 1997a). But – as argued – there is no a priori reason to emphasise this particular spatial scale, irrespective of time and place. A strong case is made for the importance of the subnational scale as an appropriate mode of analysis. Localised input-output relations between the actors of the system, knowledge spillovers and their untraded interdependencies lie at the heart of this reasoning. The concluding part of the paper summarises some of the major findings of the discussion and points to some directions for future research.

2   Innovation and Technological Change

The discussion that follows in this section builds on the converging elements from two different traditions of theoretical and empirical work in technological change. The first tradition – which might be labelled evolutionary or neo-Schumpeterian – has always stressed the dynamic and systemic nature of the innovation and diffusion processes, as well as the close links between the two. It has illustrated its theoretical framework empirically with detailed case studies as well as with work related to the history of technology. The second tradition that might be labelled neoclassical has steadily moved towards a view that emphasises dynamic and systemic aspects and, thus, shares many common elements with the first. Its rigorous theoretical and empirical framework is contributing to the clarification and quantification of many of the, as yet, unanswered questions in this field.

2.1 Innovation as an Interactive Process

Technological change is a complex process that is not yet fully understood. This complexity stems partially from the diverse set of phenomena that are subsumed under the term innovation. Bienaymé (1986), for example, distinguishes between product innovations; innovations destined to resolve, circumvent or eliminate a technical difficulty in manufacture or to improve services; innovations for the purpose of saving inputs (e.g. energy conservation, automation); and, innovations to improve the conditions of work. These very different phenomena have made generalisation difficult (Malecki 1997). For a long time thinking about technological change and innovation has been guided by linear models – in the 1950s and 1960s by the technology-push and then the need-pull model. In the former, the development, production and marketing of new technology (defined in the sense of Mansfield et al. 1982 as consisting of a pool or set of knowledge) was assumed to follow a well defined time sequence which began with basic and applied research activities, involved a product development stage, and then led to production and possibly commercialisation. In the second model, this linear sequential process emphasised demand and markets as the source of ideas for R&D activities. These models that have guided the formulation of national R&D policies in the past have come under increasing attack in recent years for several reasons, not the least of which is the absence of feedback loops between the downstream (market related) and upstream (technology related) phases of innovation. Intensifying competition and shorter product life cycles are necessitating a closer integration of R&D with the other phases of the innovation process.

This criticism has led to a broader view of the process of innovation as an interactive process. The presently emerging innovation theory emphasises the central role of feedback effects between the downstream and upstream phases of innovation and the numerous interactions between science, technology and innovation related activities within and among firms. Through interaction and feedback different pieces of knowledge become combined in new ways or new knowledge is created.

Figure 1 represents what is referred to as the chain-linked model (Kline and Rosenberg 1986, OECD 1992, Malecki 1997). The innovation process at the firm level is portrayed as a set of activities that are linked to one another through complex feedback loops. The process is visualised as a chain, starting with the perception of a new market opportunity and/or a new invention based on novel pieces of scientific and/or technological knowledge followed by the analytical design for a new product or process and testing, redesign and production, and distribution and marketing. Short feedback loops link each downstream phase in the central chain with the phase immediately preceding it. Longer feedback loops link perceived market demand and product users with phases upstream. The second set of relationships visualised in Figure 1 link the innovation process embedded in the firm with its firm specific knowledge base, the general scientific and technological pool and with research in general.

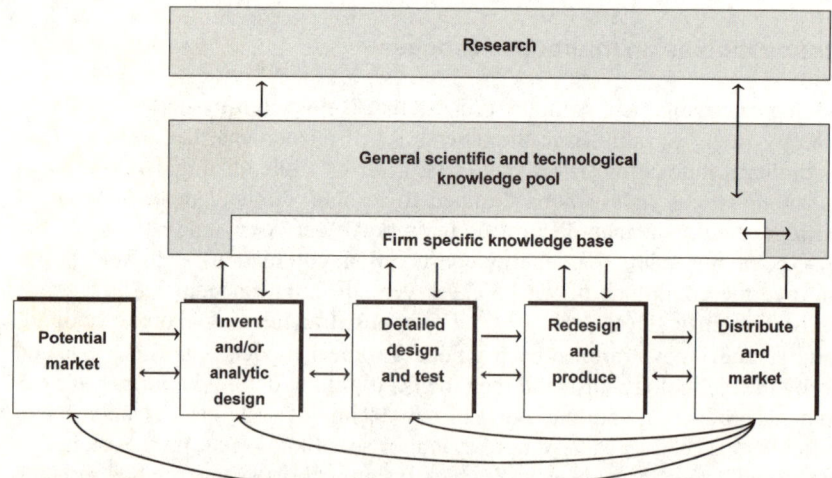

Figure 1   An interactive model of the innovation process: Feedbacks and interactions (Fischer 1999) Source: Adapted with minor changes from Kline and Rosenberg (1986), Myers and Rosenbloom (1996), Malecki (1997)

The model combines two types of interaction. One concerns processes that occur through new forms of product development practice within the firm and create appropriate feedback relationships (see, for example, Nonaka and Takeuchi 1995). The second refers to relationships external to a given firm with other companies such as customers, suppliers of inputs (including finance and knowledge), research institutions and even competitors. Co-operation can take place with various mixes of internal and external actors. Under this model technological innovation is seen to be the result of a complex interplay among various actors, with partly common and partly conflicting interests. Technological progress is, thus, dependent on how the actors interact with each other, internally and externally.

In recent years, new forms of interfirm agreements bearing on technology have developed alongside the traditional means of technology transfer – licensing and trade in patents – and they often have become the most important way for firms, regions and nation states to gain access to new knowledge and key technologies. The network form of governance – involving both action by mediating third parties as well as 'relational contracting' between firms – can overcome market imperfections as well as the rigidities of vertically integrated hierarchies. The limitations of these two modes of transaction in the context of knowledge and innovation diffusion have pushed interfirm agreements to the forefront of corporate strategy in the last few decades (Chesnais 1988).

There are many definitions of *innovation networks* (see De Bresson and Amesse 1991; Freeman 1991); however, the one offered by Tijssen (1998, p. 792) captures the most important features of the network mode. He suggests defining a network as

> 'an evolving mutual dependency system based on resource relationships in which their systemic character is the outcome of interactions, processes, procedures and institutionalisation. Activities within such a network involve the creation, combination, exchange, transformation, absorption and exploitation of resources within a wide range of formal and informal relationships.'

In a network mode of resource allocation, transactions occur neither through discrete exchanges nor by administrative fiat, but through networks of individuals or institutions, engaged in reciprocal, preferential and supportive actions (Powell 1990).

Networks show a considerable range and variety in content, which differ according to specific circumstances. Their nature is shaped by the objectives for which network linkages are formed. For example, they may focus on a single point of the R&D-to-commercialisation process or may cover the whole innovation process. The content and shape of a network will also differ according to the nature of relationships and linkages between the various actors involved (see Chesnais 1988). At the one end of the spectrum lie highly formalised relationships. The formal structure may consist of regulations, contracts and rules that link actors and activities with varying degrees of constraint. At the other end are the network relations of a mainly informal nature, linking actors through open chains. Such relations are very hard to measure (Freeman 1991). Whenever interfirm transactions tend to be small in scale, variable and unpredictable in nature, requiring face-to-face contact, then network formation will focus on the close proximity of the partners involved (Storper 1997).

For firms, networks represent a response to quite specific circumstances. Where complementarity is a prerequisite for successful innovation, network agreements may be formed in response to specific proprietary tacit knowledge. The exchange of such complementary assets can take place only through very close contacts and personalised, and generally localised, relationships (OECD 1992). It is especially shared trust that establishes an environment which facilitates exchange of knowledge (Maskell and Malmberg 1999). When technology is moving rapidly, flexibility and reversibility along with risk sharing represent another reason for preferring a network mode. The network mode provides a far higher degree of flexibility (OECD 1992). Interfirm agreements are easier to dissolve than internal developments or mergers. Porter and Fuller (1986) stress speed as being among the advantages that networks have over acquisition or internal development through arm's length relationships. This advantage is becoming increasingly important as product life cycles have shortened and competition has intensified. High R&D costs may be another distinct reason for networking and can force management, especially in smaller firms, to pool resources with other firms, in some cases even with competitors (OECD 1992).

2.2 Knowledge Generation and Diffusion

Recognition of the interactive nature of the innovation process has resulted in the break down of the earlier distinction between innovation and diffusion. The creation of knowledge and its assimilation are part of a single process. Firms need to absorb, create and exchange knowledge interdependently. In other words, innovation and diffusion usually emerge as a result of an interactive and collective process within a web of personal and institutional connections which evolve over time.

Knowledge transfer may occur through disembodied or equipment-embodied diffusion. The latter is the process by which innovations spread in the economy through the purchase of technology-intensive machinery, such as computer-assisted equipment, components and other equipment. Disembodied technology diffusion refers to the process where technology and knowledge spread through other channels not embodied in machinery (OECD 1992). This type of knowledge transfer may occur via descriptions of new products or production processes found in catalogues, publications or patent applications, but also via seminars and conferences, and R&D personnel turnover. It can also be the byproduct of mergers and acquisitions, joint ventures or other forms of interfirm co-operation.

Two notions are central to an understanding of disembodied technology diffusion: absorption capacity and knowledge spillovers. The *absorption capacity* of firms and other organisations refers to the ability to learn, assimilate and use knowledge developed elsewhere through a process that involves substantial investments, especially of an intangible nature (Cohen and Levinthal 1989). This capacity depends crucially on learning experience, which in turn may be enhanced by in-house R&D activities. The concept of absorption capacity implies that in order to have access to a piece of knowledge developed elsewhere, it is necessary to have done R&D on something similar (Saviotti 1998). Thus, R&D may be viewed as serving a dual, but strongly interrelated role: first, developing new products and production processes, and second, enhancing the learning capacity.

Firms, especially smaller firms, that lack appropriate in-house R&D facilities have to develop and enhance their absorption capacity by other means, such as learning from customers and suppliers, interacting with other firms and taking advantage of knowledge spillovers from other firms and organisations (Lundvall 1988). These sources provide the know-why (i.e., procedural knowledge), know-how (i.e., skills and competences) and know-what (i.e., factual knowledge) important for entrepreneurial success (Johannisson 1991; Malecki 1997). Network arrangements of different kinds provide a firm the assistance necessary to take advantage of outside knowledge.

Knowledge spillovers (i.e., knowledge created by one firm can be used by another without compensation or with compensation less than the value of the knowledge) arise because knowledge and innovation are partially excludable and non-rivalrous goods (Romer 1990). Lack of *excludability* implies that knowledge producers have difficulty in fully appropriating the returns or benefits and thereby preventing other firms from utilising the knowledge without compensation (Teece 1986). Patents and other devices, such as lead times and secrecy, are a way for knowledge producers to partially capture the benefits related to their knowledge creation. It is important to recognise that even a completely codified piece of knowledge can not be utilised at zero cost by everyone. Only those economic agents who know the code are able to do so (Saviotti 1998).

By non-rivalry knowledge distinguishes itself from all other inputs in the production process. *Non-rivalry* means essentially that a new piece of knowledge can be utilised many times and in many different circumstances, for example by combining with knowledge coming from another domain. The interest of the knowledge users is, thus, best served if innovations, once produced, are widely available and diffused at the lowest possible cost. This implies an environment rich in knowledge spillovers (OECD 1992).

Recent understanding of the nature of knowledge associated with technological innovation processes is at the heart of conceptual advance. Innovation – in the form of advancing technology – combines two types of knowledge: *codified* [also termed explicit] knowledge drawn from previous experience and *uncodified* [implicit] knowledge which is industry specific, firm specific or even individual specific, and has some degree of tacitness. In each technology there are elements of tacit and specific knowledge. Following Polanyi (1966) *tacitness* refers to those elements of knowledge that persons have which are ill-defined, uncodified and which they themselves can not fully articulate and which differ from person to person, but which may to some degree be shared by collaborators who have a common experience. Most shared knowledge is seldom completely tacit or completely codified [i.e., explicit]. In most cases a piece of knowledge can be located between these two extremes. Knowledge is not created codified and is always at least partly tacit in the minds of those who create it. Codification is required because knowledge creation is a collective process that requires complex mechanisms of communication and transfer (Saviotti 1988). As tacit components – such as common practice based on modes of interpretations, perceptions and value systems – in the firm's knowledge base increase, knowledge accumulation becomes more experience based, i.e., based on firm specific skills and competences like reliability and reputation. Such forms of knowledge can only be shared, communicated or transferred through network types of relationships.

In an economic system where innovation is crucial for competitiveness, the organisational ability to create knowledge becomes the foundation of innovating firms. Nonaka and Takeuchi (1995) have recently proposed a simple, but elegant model to account for the generation of knowledge in the firm. What they label the knowledge-creating company is based on the organisational interaction between codified (explicit) knowledge and implicit knowledge at the source of innovation. Organisational knowledge creation that reflects the importance of institutional learning processes involves two forms of interactions: between tacit and explicit knowledge, and between individuals and the organisation. The interaction between the two forms of knowledge is central to the dynamics of knowledge creation in the business organisation. It will bring about four major processes of knowledge conversion that require special learning processes and all together constitute knowledge creation (see Figure 2):

- *from tacit into explicit knowledge*, the externalisation mode that holds the key to knowledge creation because it generates new explicit concepts from tacit knowledge; codification is at the heart of this mode,

- *from explicit to tacit knowledge*, the internalisation mode that is closely related to learning-by-doing and leads to operational/procedural knowledge,

- *from tacit into tacit knowledge*, the socialisation mode that is a procss of sharing experiences and thereby creating some sort of novel tacit knowledge such as, e.g., technical skills;

- *from explicit to explicit knowledge*, the combination mode that is a process which involves combining different bodies of explicit knowledge in order to create systemic knowledge; a mode that is widely occurring in instructing, training and supervision of employees.

It is important to note that knowledge is performed by individuals, not by the organisation itself. If the knowledge can not be shared with others or is not amplified at the group level the knowledge does not move up to the organisational level.

Tacit Knowledge To Explicit Knowledge

Tacit Knowledge	Sympathised Knowledge *Socialisation*	Conceptual Knowledge *Externalisation*
From		
Explicit Knowledge	Procedural Knowledge *Internalisation*	Systemic Knowledge *Combination*

Figure 2   Four major processes of knowledge conversion (Nonaka and Takeuchi 1995)

The core of the organisational knowledge creation process, Nonaka and Takeuchi (1995) argue, takes place at the group level, but the organisation provides the necessary enabling conditions, the organisational context formed by conventions, managerial ideologies, customs, habits and established business practices that facilitate the creation and accumulation of knowledge at the organisational level. Organisational knowledge creation is, thus, a complex non-linear interactive process characterised by a continuous and dynamic interaction between tacit and explicit forms of knowledge that is shaped by shifts between the above four different modes of knowledge transformation. This organisational knowledge creation process requires the participation of the workers in the innovation process so that they do not keep their tacit knowledge solely for their own profit. It also requires stability of the labour force in the firm because only then it is rational for the individual to transfer his/her knowledge to the organisation, and for the firm to diffuse explicit knowledge (Castells 1996). On-line communications along with artificial agents and expert systems have become powerful tools in recent times in assisting to manage the complexity of necessary organisational links in the knowledge creation process.

3 Systems of Innovation: Conceptual Framework and Localised Systems

Actions that aim at favouring the building and consolidation of innovation related networks have always been an implicit component of national science and technology policies. Today, it seems possible to approach this dimension of policy making more explicitly on the basis of the systems of innovation approach, a conceptual framework that has recently received considerable attention both in academia and in the policy arena (Freeman 1987, Lundvall 1992, Nelson 1993, OECD 1994, Edquist 1997a).

3.1 A Conceptual Framework to Innovation Systems Research

The systems of innovation approach is not a formal theory, but a conceptual framework – a framework in its early stage of development. The idea that lies at the centre of this framework is – as already mentioned above – that the economic performance of territories (regions or countries) depends not only on how business corporations perform, but also on how they interact with each other and with the public sector in knowledge creation and dissemination. Innovating firms operate within a common institutional set-up, and they jointly depend on, contribute to and use a common knowledge infrastructure. Consequently, the approach places innovation, knowledge creation and diffusion at its very centre. Innovation and knowledge creation are viewed as interactive and cumulative processes contingent on the institutional set-up. It departs from the network school of research (Håkansson 1987) with its emphasis on the role of the institutional set-up, i.e., that institutions play in the innovation process (see Edquist and Johnson 1997). The concept of institutions refers at an abstract level to the recurrent patterns of behaviour: socially inherited habits, conventions including regulation, values and routines (Morgan 1997) that assist in regulating the relations between people and groups of people within as well as between and outside the organisations.

A *system of innovation* can be thought of as consisting of a set of actors or entities such as firms, other organisations and institutions that interact in the generation, use and diffusion of new – and economically useful – knowledge in the production process. At the current stage of development there is no general agreement as to which elements and relations are essential to the conceptual core of the framework and what is their precise content (Edquist 1997b). This leaves room for a conceptual discussion.

Systems that attempt to encompass the whole innovation process may be expected to include four key building blocks that comprise groups of actors sharing some common characteristics and institutions governing the relations within and between the groups (see Figure 3):

- The *manufacturing sector*. This sector is made up of manufacturing firms [the central actors in the system of innovation] and their R&D laboratories that play a fundamental role in performing research and technological development.

- The *scientific sector*. The scientific sector plays a very important role in technological innovation. It consists of two components: a training component that includes educational and training organisations on which the supply of scientists, engineers, technicians and other skilled workers possessing appropriate skill profiles depends, and a research component including universities and other research organisations that generate and diffuse knowledge and produce documents in the form of scientific publications. This sector involves those agents [government, private non-profit, universities, higher education] that both fund and carry out research or offer education.

- The *sector of producer services*. This sector includes organisations or units within larger organisations which provide assistance or support to industrial firms for the development and/or introduction of new products or processes. This may take any of the following forms: financial, technical advice or expertise, physical (equipment, software, computing facilities), marketing or training related to new technologies or procedures.

- The *institutional sector*. Many of the tasks that a typical firm must perform require co-ordination, either within the firm between various groups of employees or outside it with other suppliers, other firms, and providers of producer services including finance. There is a variety of ways in which the performance of these tasks can be co-ordinated, each involving different kinds of firm behaviour. But in general one can distinguish market co-ordination that relies on the kind of market institutions neo-classical economics usually assumes to be important, and non-market co-ordination that utilises a greater range of institutional arrangements. The latter depends upon the presence of formal and informal institutions that regulate the relations between the actors of the system, enhance their innovation capacities and manage conflicts and co-operation. Two types of such institutions may be distinguished (see, for example, Edquist and Johnson 1997): formal institutions including, e.g., employer associations, legal and regulatory frameworks, and informal institutions including the prevailing set of rules, conventions and norms that prescribe behavioural roles and shape expectations.

To describe and compare systems of innovation in the broad sense one has to open the boxes of the subsystems, identify the constituent elements and specify those relations between and within the subsystems that have importance for innovation performance. A first source of diversity among different systems might be due to differences in the macroeconomic context, the quality of information and communication infrastructures as well as in factor and product market conditions.

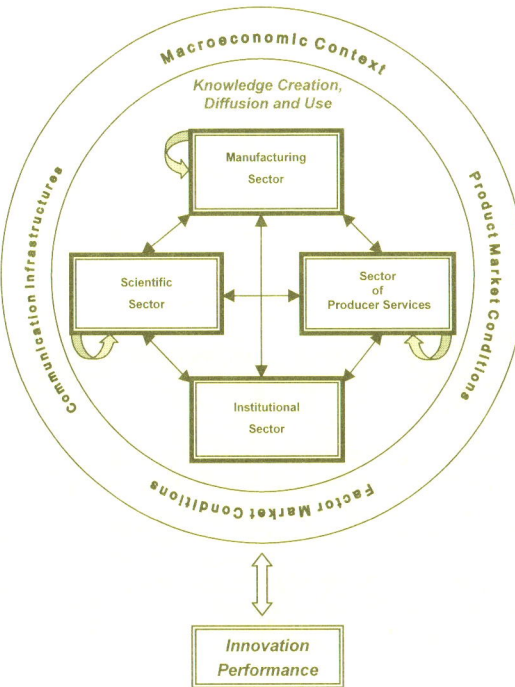

Figure 3   The major building blocks of a system of innovation

The innovation performance of an economy is notably determined by the characteristics and abilities of its individual firms and other organisations contingent on its institutions, but also very much by the different kinds of relations between them, i.e., the ways they interact with each other and with the sector of institutions. The character as well as the change of these interaction patterns are central aspects of innovation systems. Linkages within and between the sectors can be specified in terms of knowledge and information flows, flows of investment funding, flows of authority and labour mobility (scientists, technicians, engineers and other skilled workers) as important mechanisms for the transfer of tacit forms of knowledge particularly from the scientific to the manufacturing sector, but also within the latter.

   Network analysis may assist to identify the central actors in the four subsystems (building blocks) in specific cases, and of the type of information and knowledge they exchange. Different kinds of norms, conventions and established practices that are expected to have important implications for knowledge creation and learning are forming the economy's patterns of interaction, both inside firms and other organisations and between them. Searching for and explaining interaction patterns that lead to the creation, dissemination and use of knowledge is part of the systems of innovation approach (Johnson 1997). It can be hypothesised, for example, that there will be strong and weak, regular and irregular interactions which shape the system.

Firms are the main carriers of technological innovation. Their capacity to innovate is partly determined by their own capabilities, and partly by their absorption capacities. Increasing complexity, costs and risks in innovation enhance the role of collaboration and networking in the innovation process to reduce moral hazard and transaction costs. In addition to traditional market-mediated relations such as the purchase of equipment and licensing of technology, firms exchange information and engage in mutual learning in their roles as customers, suppliers and subcontractors, and even competitors. A coherent system of innovation has necessarily to include a series of more or less co-ordinated network-like relations such as (Fischer 1999):

- *Customer-producer relations,* i.e., forward linkages of manufacturing firms with distributors, value-added resellers and end users,

- *Producer-manufacturing supplier relations* which include subcontracting arrangements between a client and its manufacturing suppliers of intermediate production units,

- *Producer-service supplier relations* which include arrangements between a client and its producer service partners (especially computer and related service firms, technical consultants, business and management consultants),

- *Producer network relations* which include all co-production arrangements (bearing on some degree or another on technology) that enable competing producers to pool their production capacities, financial and human resources in order to broaden their product portfolios and geographic coverage,

- *Science-industry collaboration* between universities and industrial firms at various levels pursued to gain rapid access to new scientific and technological knowledge and to benefit from economies of scale in joint R&D, such as direct interactions between particular firms and particular faculty members, or joint research projects, as through consulting arrangements, or mechanisms that tie university or research programs to groups of firms.

3.2 Territorially Based Systems of Innovation – National or Regional?

Within the systems of innovation approach to innovation analysis, different types of systems have been defined. A major distinction can be made (see Gregersen and Johnson 1997) between

- *sectoral* or *technological systems* that are based on the concept of technological regimes and take a specific sector or a specific technology as their point of departure (see, for example, Carlsson 1995, Breschi and Malerba 1997), and

- *localised* (or territorially based) *systems* which – built on some kind of spatial proximity – may be manifested at different geographical scales – as either *local, regional* (i.e., subnational), *national* or *global systems of innovation* (see, for example, Lundvall 1992, Nelson 1993, Braczyk et al. 1998, Malecki and Oinas 1999).

Whether a system of innovation should be sectorally/technologically or spatially defined depends on the objective and context of a study at hand. These two basic variants of the systems of innovation approach complement rather than exclude each other.

Geographical proximity can be considered as a *necessary, but not sufficient* precondition for the existence of a territorially based system of innovation. A proximity that is only geographic in nature can provide the basis for the presence of an agglomeration of firms, but not necessarily for the presence of a system of innovation. The potential of an innovation system crucially depends, above all else as discussed above, on two factors: geographic proximity and technological proximity. *Geographic proximity* indicates the positioning of actors within a given spatial framework, while *technological proximity* pertains to the association with the set of vertical or horizontal interdependencies within the scope of production relationships. The transformation of these two types of proximity into a territorially based system of innovation assumes that they be institutionally organised and structured (Kirat and Lung 1999). Thus, territorially defined systems of innovation are grounded in collective action at a territorial level. The cohesiveness of a territorially based system of innovation is provided by a spectrum of informal institutions, i.e., the territorially prevailing set of rules, conventions and norms (Kirat and Lung 1999).

The concept of territorially based systems of innovation evolved first in a national context (Freeman 1987), and then in a regional context (see, for example, Cooke et al. 1997, Brazcyk et al. 1998, Malecki and Oinas 1999). The tradition of studying national systems of innovation has been a recent development (see, for example, Lundvall 1992, Nelson 1993, Niosi et al. 1993, OECD 1994, Edquist 1997a). Interesting questions and findings have emerged from this literature that sought to establish the extent of convergence and divergence among national innovation systems. This question was of special interest in Europe, given the

emergence of European innovation related institutions that have developed simultaneously with European Community institutions (see Caracostas and Soete 1997).

It is increasingly being recognised that important elements of the process of
innovation become transnational and global, or regional rather than national. The
driving forces behind this recognition are two processes that are simultaneously at
work today: the process of globalisation of factor and commodity markets and the
regionalisation of knowledge creation and learning. This concurs with the view
expressed in Ohmae's work on the hollowing-out of the nation state in an
increasingly borderless economic world and its identification of the regional rather
than the national scale as the relevant economic scale at which leading edge
business competitiveness is being organised in practical terms (Ohmae 1995).
Regions like Baden-Württemberg, Wales, Hongkong-Canton are conceived as
much more economically meaningful than, for example, Italy with its abiding
north/south divide (Brazcyk et al. 1998).

This awareness does not claim that the national scale is unimportant or
irrelevant. This scale continues to be crucial in some circumstances. But it is
becoming increasingly clear that there is no a priori reason to privilege this
particular spatial scale in systems of innovation research, irrespective of time and
place (see also Hudson 1999).

A strong case is made today that the regional [i.e., subnational] scale is growing
in importance as a mode for innovation systems research. The main argument for
this is that regional agglomeration provides the best context for an innovation
based learning economy (Hudson 1999), for knowledge creation and diffusion and
learning. Specific forms of knowledge creation, especially the tacit forms, and of
technological learning are both localised and territorially specific. The firms that
master knowledge that is not fully codifiable are tied into various kinds of
networks with other firms and organisations through localised input-output
relations, knowledge spillovers and their untraded interdependencies (Storper
1997). In some cases market exchange, knowledge spillovers and untraded
relations are woven between the various activities within the scope of vertical or
horizontal production relationships, but often they are separated.

Formal exchange (i.e., traded interdependencies) and – more importantly –
knowledge spillovers and their untraded interdependencies lie at the heart of this
line of reasoning:

- *First*, localised input-output relations constitute webs of customer-producer and
  producer-supplier relations that are essential to communicate information about
  both technological opportunities and user needs. The user/supplier and
  producer will gradually develop a common code of communication, making the
  exchange of information more efficient. To leave a well-established user-
  producer or producer-supplier relationship becomes increasingly costly and
  involves a loss of information capital (Lundvall 1992).

- *Second*, knowledge spillovers occur because knowledge created by one firm or
  another organisation is typically not contained within that organisation, and

thereby creates value for other firms and other firms' customers. Knowledge spillovers are especially likely to result from basic research, but they are also generated from applied research and technological development. This can occur, for example, in obvious ways such as reverse engineering of products, but also in less obvious ones such as when one firm's abandonment of a particular research line signals to others that the line is unproductive and, thus, saves them the expense of learning this themselves. The spillover beneficiary may use the new knowledge to copy or imitate the commercial products or processes of the innovator, or may use the knowledge as an input to R&D leading to other new products or processes. Three vehicles of such spillovers may be distinguished: first, the scientific sector with its general scientific and technological knowledge pool; second, the firm specific knowledge pool; and, third, the business-business and university-industry relations that make them possible. Once the central role of knowledge spillovers is recognised, a place for informal institutions appears.

- *Third*, untraded interdependencies or regional assets are less tangible benefits that attach to the process of economic co-ordination and organisational knowledge creation. They are derived from geographical clustering, both economic – such as the development of a pooled labour market – and sociocultural – such as developed routines, shared values, norms, rules and trust that facilitate interactive processes and mutual understanding in the transmission of information and knowledge. Because tacit knowledge is collective in nature and wedded to its sociocultural context, it is more territorially and place specific than is generally thought.

Thus, from a more general perspective it can be argued that it is the combination of territorially embedded Marshallian agglomeration economies, knowledge creation and spillovers and their untraded interdependencies that defines the importance of the regional scale in innovation systems research.

4   Summary and Outlook

The diffusion of the systems innovation approach – in different versions and variations – has been surprisingly fast in academic circles, and is also very much used in a national context by national governments as well as supranational organisations like the European Union and the Organisation for Economic Co-operation and Development. The approach seems to be very attractive to policy makers who look for frameworks to understand differences between national and regional economies and various ways to support technological change. The attractivity stems from three basic characteristics of the approach that deserve to be summarised here:

- First, it places innovations and knowledge creation at the very centre of focus, and goes beyond a narrow view of innovation to emphasise the interactive and dynamic nature of innovation.
- Second, it represents a considerable advance over the network school of innovation (Håkansson 1987) by a decisive shift in focus from firm to territory, from the knowledge-creating firm to the knowledge-creating territory.

- Third, it views innovation as a social process that is institutionally embedded, and, thus, puts special emphasis on the institutional context and forms [i.e., formal and informal institutions] through which the processes of knowledge creation and dissemination occur.

Adoption of this approach overcomes the weaknesses of case studies because a common conceptual framework is used. Its advantage is that it allows for a systematic comparison of innovation activities in different localised systems. Conducting comparative studies can lead to the identification of functional communalities as well as to the discovery of specific and generic problems within the innovation process. Three types of innovation analysis may be performed, depending on the context of analysis:

- the first refers to the *micro-level of the system* and attempts to analyse the internal capabilities of selected firms and the links surrounding them [knowledge relationships with other firms and with non-market institutions] with the purpose to identify unsatisfactory or problematic links in the value chain;

- the second refers to the *meso-level of the system* and focuses on specific subsystems and attempts to map knowledge and other interactions within and between subsystems. This may involve the measurement of various types of knowledge flows: interactions among manufacturing firms; interactions between manufacturing firms and universities including joint research, co-patenting, co-publications and more informal relations; interactions between manufacturing firms and other innovation supporting units such as innovation funding; and personnel mobility focusing on the movement of scientific and technical personnel within the enterprise sector and between the scientific and the enterprise sector;

- the third refers to the *macro-level of the system* and typically involves the use of macro-indicators such as, for example, R&D personnel ratios, R&D expenditure intensity rates, innovation rates, patent intensity rates, networking indicators of various kinds to characterise the system at hand in general terms.

In concluding it should be stressed that many of the fundamental ideas in the approach discussed in this paper still lack a firm and more rigid conceptual foundation, and definitional and conceptual dimensions are important. The topics include the character and changeableness of a system of innovation's knowledge base and its dependence on specific innovation infrastructures, as well as the nature and importance of formal and especially informal institutions, technologies and territories as levels of analysis (Johnson 1997). There are several unresolved problems, for example, of how to define and describe the structure and change of the institutional set-up and to connect it to innovation. Different kinds of rules, norms, conventions and shared practices that form regional and national economy's patterns of interaction have important implications for the creation, application and dissemination of knowledge. It follows from this that there is a need for much more specific conceptual categories, sharp enough to tightly guide empirical work. Empirical research still often lacks the ability to dig into the specifics of hard-to-measure issues, such as trust-building, coalition-building, control relations or culturally loaded industrial practices (Malecki et al. 1999).

Another line for future research efforts refers to the complex interplay of processes taking place at different spatial scales. We need to bear in mind that spatial scales are not independent entities. Individual production/commodity chains can be viewed as vertically organised structures that operate across increasingly extensive geographical scales. Cutting across these vertical structures are the territorially based systems that are manifested at different spatial scales. It is at the points of intersection of these dimensions in geographic space where specific outcomes occur and the problems of existing within globalising economy have to be resolved (Dicken 1998). This, however, means that there is a need to acknowledge the importance of external linkages, network relations, and non-market connections as one of the determining factors to economic success and to move beyond the simple statement that localised innovation systems are open systems.

References

Bienaymé A. (1986): The dynamics of innovation, *International Journal of Technology Management* 1, 133-159

Braczyk H.-J., Cooke P. and Heidenreich M. (eds.) (1998): *Regional Innovation Systems. The Role of Governances in a Globalised World*, UCL Press, London

Breschi S. and Malerba F. (1997): Sectoral innovation systems: Technological regimes, Schumpeterian dynamics, and spatial boundaries. In: Edquist C. (ed.) *Systems of Innovation. Technologies, Institutions and Organisations*, Pinter, London, pp. 130-156

Caracostas P. and Soete L. (1997): The building of cross-border institutions in Europe: Towards a European system of innovation? In: Edquist C. (ed.) *Systems of Innovation. Technologies, Institutions and Organisations*, Pinter, London, pp. 395-419

Carlsson B. (ed.) (1995): *Technological Systems and Economic Performance: The Case of Factory Automation*, Kluwer, Dordrecht, Boston

Castells M. (1996): *The Rise of the Network Society*, Blackwell, Oxford [UK], Malden [MA]

Chesnais F. (1988): Technical cooperation agreements between firms, *STI Review* 4, 51-120

Cohen W.M. and Levinthal D.A. (1989): Innovation and learning: The two faces of R&D, *Economic Journal* 99 (397), 569-596

Cooke P. (1998): Introduction: Origins of the concept. In: Braczyk H.-J., Cooke P. and Heidenreich M. (eds.) *Regional Innovation Systems. The Role of Governances in a Globalised World*, UCL Press, London, pp. 2-25

Cooke P., Gomez Uranga M. and Etxebarria G. (1997): Regional innovation systems: Institutional and organisational dimensions, *Research Policy* 26 (4/5), 475-491

DeBresson C. and Amesse F. (1991): Networks of innovators: A review and introduction to the issue, *Research Policy* 20 (5), 363-379

Dicken P. (1998): *Global Shift. Transforming the World Economy. Third Edition*, The Guilford Press, New York, London

Edquist C. (1997a): *Systems of Innovation. Technologies, Institutions and Organisations*, Pinter, London

Edquist C. (1997b): Systems of innovation approaches – their emergence and characteristics. In: Edquist C. (ed.) *Systems of Innovation. Technologies, Institutions and Organisations*, Pinter, London, pp. 1-35

Edquist C. and Johnson B. (1997): Institutions and organisations in systems of innovation. In: Edquist C. (ed.): *Systems of Innovation. Technologies, Institutions and Organisations*, Pinter, London, pp. 41-63

Fischer M.M. (1999): The innovation process and network activities of manufacturing firms. In: Fischer M.M., Suarez-Villa L. and Steiner M. (eds.) *Innovation, Networks and Localities*, Springer, Berlin, Heidelberg, New York, pp. 11-27

Freeman C. (1991): Networks of innovators: A synthesis of research issues, *Research Policy* 20 (5), 499-514

Freeman C. (1987): *Technology and Economic Performance: Lessons from Japan*, Pinter, London

Gregersen B. and Johnson B. (1997): Learning economies, innovation systems and European integration, *Regional Studies* 31 (5), 479-490

Håkansson H. (1987): *Industrial Technological Development: A Network Approach*, Croom Helm, London

Hudson R. (1999): The learning economy, the learning firm and the learning region: A sympathetic critique of the limits to learning, *European Urban and Regional Studies* 6 (1), 59-72

Johannisson B. (1991): University training for entrepreneurship: Swedish approaches, *Entrepreneurship and Regional Development* 3, 67-82

Johnson B. (1997): Systems of innovation: Overview and basic concepts. In: Edquist C. (ed.) *Systems of Innovation. Technologies, Institutions and Organisations*, Pinter, London, pp. 36-40

Kirat T. and Lung Y. (1999): Innovations and proximity. Territories as loci of collective learning processes, *European Urban and Regional Studies* 6 (1), 27-38

Kline S.J. and Rosenberg N. (1986): An overview of innovation. In: Landau R. and Rosenberg N. (eds.) *The Positive Sum Strategy*, National Academy Press, Washington, pp. 275-305

Lundvall B.-Å. (ed.) (1992): *National Systems of Innovation: Towards a Theory of Innovation and Interactive Learning*, Pinter, London

Lundvall B.-Å. (1988): Innovation as an interactive process: From user-producer interaction to the national system of innovations. In: Dosi G., Freeman C., Nelson R.R., Silverberg G. and Soete L. (eds.) *Technical Change and Economic Theory*, Pinter, London, pp. 349-369

Malecki E.J. (1997): *Technology & Economic Development. The Dynamics of Local, Regional and National Competitiveness*, Longman, Harlow, Essex

Malecki E.J. and Oinas P. (eds.) (1999): *Making Connections. Technological Learning and Regional Economic Change*, Ashgate, Aldershot

Malecki E.J., Oinas P. and Ock Park S. (1999): On Technology and Development. In: Malecki E.J. and Oinas P. (eds.) *Making Connections. Technological Learning and Regional Economic Change*, Ashgate, Aldershot, pp. 261-275

Mansfield E., Romeo A., Schwartz M., Teece D.J., Wagner S. and Brach P. (1982): *Technology Transfer, Productivity, and Economic Policy*, W.W. Norton, New York

Maskell P. and Malmberg A. (1999): The competitiveness of firms and regions. 'Ubiquitification' and the importance of localized learning, *European Urban and Regional Studies* 6 (1), 9-25

Morgan K. (1997): The learning region: Institutions, innovation and regional renewal, *Regional Studies* 31 (5), 491-503

Myers M.B. and Rosenbloom R.S. (1996): Rethinking the role of industrial research. In: Rosenbloom R.S. and Spencer W.J. (eds.) *Engines of Innovation: US Industrial Research at the End of an Era*, Harvard Business School Press, Boston, pp. 209-228

Nelson R.R. (ed.) (1993): *National Innovation Systems. A Comparative Analysis*, Oxford University Press, New York, Oxford [UK]

Noisi J., Saviotti P., Bellon B. and Crow M. (1993): National systems of innovation. In search of a workable concept, *Technology in Society* 15 (2), 207-227

Nonaka I. and Takeuchi H. (1995): *The Knowledge-Creating Company. How Japanese Companies Create the Dynamics of Innovation*, Oxford University Press, New York, Oxford [UK]

OECD (1994): National Systems of Innovation: General Conceptual Framework. DSTI/STP/TIP 94(4), Organisation for Economic Co-operation and Development, Organisation for Economic Co-operation and Development, Paris

OECD (1992): *Technology and Economy: The Key Relationships*, Organisation for Economic Co-operation and Development, Paris

Ohmae K. (1995): *The End of the Nation State*, Free Press, New York

Polanyi M. (1966): *The Tacit Dimension*, Routledge & Kegan Paul, London

Porter M.E. and Fuller M.B. (1986): Coalitions and global strategy. In: Porter M.E. (ed.) *Competition in Global Industries*, Harvard Business School Press, Boston, pp. 315-343

Powell W.W. (1990): Neither market nor hierarchy: Network forms of organisation. In: Staw B.M. and Cummings L.L. (eds.) *Research in Organisational Behavior*, JAI Press, Greenwich [CT] pp. 295-335

Romer P. (1990): Endogenous technical progress, *Journal of Political Economy* 98 (5) Part 2, 71-103

Saviotti P.P. (1998): On the dynamics of appropriability, of tacit and of codified knowledge, *Research Policy* 26 (7/8), 843-856

Saviotti P.P. (1988): Information, entropy and variety in technoeconomic development, *Research Policy* 17 (2), 89-103

Storper M. (1997): *The Regional World. Territorial Development in a Global World*, The Guilford Press, New York, London

Teece D.J. (1986): Profiting from technological innovation: Implications for integration, collaboration, licensing and public policy, *Research Policy* 15 (6), 285-305

Tijssen R.J.W. (1998): Quantitative assessment of large heterogeneous R&D networks: The case of process engineering in the Netherlands, *Research Policy* 26 (7/8), 791-809

9 The Role of Space in the Creation of Knowledge in Austria: An Exploratory Spatial Data Analysis

with *J. Fröhlich, H. Gassler and A. Varga*

The relationship between knowledge spillovers and space is extremely complex and, at the current state of research, only partially understood. This is partly due to the fact that knowledge spillovers are difficult to measure. The chapter makes a modest attempt to shed some light on the role of space in the creation of technological knowledge in Austria. The study is exploratory rather than explanatory in nature and based on descriptive and exploratory techniques such as Moran's I test for spatial autocorrelation and the Moran scatterplot. Clusters of the output of the knowledge creation process (measured in terms of patent counts) are compared with spatial concentration patterns of two input measures of knowledge production: private R&D and academic research. In addition, employment in manufacturing is considered to capture agglomeration economies. The analysis is based on data aggregated for two-digit ISIC industries and at the level of Austrian political districts. It explores the extent to which knowledge spillovers are mediated by spatial proximity in Austria. A time-space comparison makes it possible to study whether divergence or convergence processes in knowledge creation have occurred in the past two decades. As in the case of any exploratory data analysis, the findings need to be treated with caution and should be viewed only as an initial pre-modelling stage for Chapter 11.

1 Introduction

Knowledge spillovers occur because knowledge created by a firm or other organisation is not normally contained solely within that organisation, but is also exploited by other firms. The spillover beneficiary may use the new knowledge to copy or imitate the commercial products of the innovator, or may use it as an input to R&D leading to the development of other new products or processes. Three vehicles of such spillovers may be distinguished: first, the scientific sector with its general scientific and technological knowledge pool; second, the firm specific knowledge pool; and, third, the business-business and university-industry relations that make them possible (Fischer 2001a).

In this chapter we make a modest attempt to shed some light on the role of space in the creation of technological knowledge in Austria. The study is exploratory rather than explanatory in nature and is based on descriptive and exploratory techniques such as *Moran's I* test for spatial autocorrelation and the Moran scatterplot. Clusters of the output of the knowledge creation process [measured by patent counts] are compared with spatial concentration patterns of two input measures of regional knowledge production: private R&D and academic research. In addition, we consider employment in manufacturing to capture agglomeration economies. The analysis is based on data aggregated by two-digit ISIC industries and at the level of Austrian political districts to explore the extent to which knowledge spillovers are mediated by spatial proximity in Austria. A time-space comparison will make it possible to study whether divergence or convergence processes in knowledge creation have occurred between the years of 1982 and 1998.

The remainder of the chapter is structured as follows. Section 2 introduces the exploratory tools used to analyse the spatial data and describes the data on which the study is based. Section 3 focuses on the identification of spatial clustering patterns of knowledge production in the last two decades, while Section 4 relates spatial distribution of knowledge inputs to spatial patterns of knowledge output. The final section summarises the research findings and points to directions for future research.

2   Methodology and Data

This contribution builds on the proposition that spatial clustering of knowledge production is induced by geographically bounded knowledge externalities: the larger the intensity of knowledge spillovers among the actors of a spatial [national, regional or local] innovation system, the higher the degree of spatial clustering of knowledge production. In order to shed some light into this issue, we have used the normalised Herfindahl index first to measure the degree of spatial concentration of both some input and output measures of knowledge production utilising political districts as the basic spatial units of analysis.

This index is defined as $HI = 1 + \ln \Sigma_i S_i^2 / \ln n$, where S_i stands for the share of the measurement of the variable of interest in basic spatial unit i of the national total and n denotes the number of basic spatial units. A major advantage of this index is that it can provide a basis for straightforward comparisons as it ranges between 0 and 1. The index takes the value of 0 if the variable of interest is evenly distributed across regions and the value of 1 if it is completely concentrated in one basic spatial unit.

Spatial autocorrelation [also referred to as spatial dependence or spatial association] in the data can be a serious problem, rendering conventional statistical analysis tools invalid and hence requiring specialised spatial analytical tools. This problem occurs in situations where the observations are not independent over space, that is where nearby basic spatial units are associated in some way. Sometimes, this association is due to a poor match between the spatial extent of the phenomenon of interest such as knowledge production in the current context and the administrative units for which data are available. Sometimes, it is due to a spatial spillover effect. The complications are similar to those found in time series analysis, but are exacerbated by the multi-directional, two-dimensional nature of dependence in space rather than the uni-directional nature of time dependence. Avoiding the pitfalls arising from spatially correlated data is crucial to good spatial data analysis (Fischer 1998, 2001b).

Exploratory analysis of area data involves identifying and describing different forms of spatial variation in the data. In the context of this contribution, special attention has been given to measuring the spatial association between observations for one variable. The presence of spatial association can be identified in a number of ways: a rigorous method is to use an appropriate spatial autocorrelation statistic, a more informal one is to use, for example, a scatterplot and to plot each value against the mean of the neighbouring areas. In the former approach to spatial autocorrelation the overall pattern of dependence in the data is summarised into a single indicator, such as *Moran's I* or *Geary's c*. Both of these require the choice of a spatial weights matrix [also referred to as contiguity matrix] that represents the topology or spatial arrangement of the data and manifests our understanding of spatial association.

In this current study Moran's I statistic is used. Moran's I is based on cross-products to measure value association:

$$I = \frac{n}{S_0} \sum_i \sum_j w_{ij}(x_i - \mu)(x_j - \mu)\left[\sum_i (x_i - \mu)^2\right]^{-1} \tag{1}$$

where n stands for the number of observations, x_i denotes an observation on a variable x at location i, w_{ij} is an element of the spatial weights matrix ($i = 1, …, n$; $j = 1, …, n$), μ the mean of the x variable, and S_0 the normalising factor equal to the sum of the elements of the weights matrix:

$$S_0 = \sum_i \sum_j w_{ij}. \tag{2}$$

For a row-standardised spatial weights matrix – the preferred way to implement this test – the normalising factor S_0 equals n (since each row sums to 1), and the statistic simplifies the ratio of a spatial cross-product to a variance. The neighbourhood or contiguity structure of a data set is formalised in a spatial weights matrix $(w_{ij}) = W$ of dimension equal to the number of observations (n), in which each row and matching column correspond to an observation pair (i, j). The elements w_{ij} of the weights in the matrix W take on a non-zero value (one for a binary matrix, or any other positive value for general weights based on the distance view of spatial association) when observations i and j are considered to be neighbours, and a zero value otherwise. By convention, the diagonal elements of the weights matrix, (w_{ij}), are set to zero. Note that the row-standardised weights matrix is likely to become asymmetric, even though the original matrix may have been symmetric.

Tests for spatial autocorrelation for a single variable in a cross-sectional data set are based in this study on the magnitude of Moran's I that combines the value observed at each basic spatial unit with the values at neighbouring locations. Basically, Moran's I is a measure of the similarity between association in value and association in space (contiguity). Spatial autocorrelation is considered to be present when the statistic for a particular map pattern has an extreme value compared to what would be expected under the null hypothesis of no spatial autocorrelation. We are interested in instances where large values are surrounded by other large values, or where small values are surrounded by other small values. This is referred to as *positive spatial autocorrelation* and implies a spatial clustering of similar values.

The exact interpretation of what is 'extreme' depends on the distribution of the test statistic under the null hypothesis, and on the chosen level of the Type I error, that is on the critical value for a given significance level. Two main approaches are used in the study to determine the distribution of a test for spatial autocorrelation under the null hypothesis. The first, and most widely used assumption, is that the data follow an uncorrelated *normal* distribution. If this is not the case, the so-called permutation approach is adopted. This utilises the data themselves to construct an artificial reference distribution by resampling the data over the basic spatial units (that is by allocating the same set of observations randomly to the different locations). The degree of 'extremeness' of the Moran I statistic for the observed pattern can then be assessed by comparing it to the frequency distribution of the random permutations. A simple rule of thumb can be based on a so-called pseudo significance level. This is computed as $(T + 1)/(M + 1)$ where T denotes the number of values in the reference distribution that are equal to or more extreme than the observed statistic, and M is the number of permutations carried out (M may be taken to be 99, for example).

Since the x-variable is in deviation from its mean, Moran's I is formally equivalent to a regression coefficient in a regression of Wx on x. The interpretation of I as a regression coefficient provides a way to visualise the linear association between x and Wx in form of a bivariate scatterplot of Wx against x, termed as a *Moran scatterplot* (Anselin 1997). The *Moran scatterplot* can be augmented with a linear regression (as a linear smoother of the scatterplot) that has Moran's I as slope, and can be used to indicate the degree of fit, the presence of outliers etc. in the usual manner. The lower left and upper right quadrants represent clustering of similar values. By contrast, the upper left and lower right quadrants contain non-clustering observations. Points in the scatterplot that are extreme with respect to the central tendency reflected by the regression slope may be outliers in the sense that they do not follow the same process of spatial dependence as the other observations. Leverage points are observations that have a large influence on the regression slope. If the regression has a positive slope (that is, positive global spatial association), points further than two standard deviations from the center $(0, 0)$ in the upper left and lower right quadrants are considered in this study as outliers. Observations that are in a two standard deviations distance from the centre in the lower left and upper right quadrants are leverage points.

The interpretation of Moran's I as a regression coefficient clearly illustrates the way in which the statistic summarises the overall pattern of linear association, in the sense that a lack of fit would indicate the presence of local pockets of non-stationarity. It also indicates that the global measure of spatial association may be a poor measure of the actual dependence in the process at hand. *Local* measures of spatial association such as the *local Moran* statistic (Anselin 1995) are suitable for detecting potential non-stationarities in a spatial data set, for example, when the spatial clustering is concentrated in one subregion of the study area only. The local Moran for an observation i may be calculated as follows:

$$I_i = (x_i - \mu)\sum_j w_{ij}(x_j - \mu) \tag{3}$$

where w_{ij} denotes the (i, j)th element of a spatial weights matrix in row-standardised form. Significant local Moran's I_i detect non-random local spatial clusters where observation i is the center of the cluster. Significance tests are based on the permutation approach (see above).

Exploratory spatial data analysis in this study focuses explicitly on the spatial aspects of both input and output measures of the knowledge production process. Given the supposedly micro scale of interactions in knowledge production, the spatial level of data aggregation should be as low as possible. Due to data availability restrictions we were forced to choose political districts as the basic units of analysis in this study. Two input measures of knowledge production are considered: R&D expenditures in manufacturing and university research expenditures. Additionally, manufacturing employment is included to proxy agglomeration effects on knowledge production in an unspecified form. Patent count data are used as indicators of knowledge output despite their widely known drawbacks and problems (Basberg 1987; Pavitt 1988; Griliches 1990; Archibugi 1992; Archibugi and Pianta 1996; Fischer et al. 1994).

Raw data on Austrian patents filed between 1982 and 1998 were provided by the Austrian Patent Office (APO). The data files contain information on the application date, name of the assignee(s), address of the assignee(s), name of the inventors(s), location of the inventor(s), one or more International Patent Classification (IPC) codes and some information on the technology field of the patent application. Since location information on the inventor(s) was not always provided, the address of the assignee was used for tracing patent activity back to the region of knowledge production. It is common in Austria that the location of both the assignee [usually the firm where the inventor has a job] and the inventor are very near, and often in the same political district. Deviation from this pattern was found only for large multiple-location companies where patent applications were submitted by the companies' headquarters. For these cases, patents were re-distributed to the addresses of the inventors when these were located in different political districts. In the case of multiple assignees located in different political districts, we followed the standard procedure of proportionate assignment. We used the MERIT concordance table (Verspagen et al. 1994) between patent classes (International Patent Classes, IPC) and industrial sectors (ISIC) to match the patent data with the two-digit ISIC codes.

Finally, we needed data on the amount of university research relevant to each two-digit ISIC industry. There are great differences in the scope and commercial applicability of university research undertaken in different scientific fields. Academic research will not necessarily result in useful knowledge for every industry, but scientific knowledge from certain academic institutes (especially those operating in the transfer sciences) is expected to be important for specific industries. To capture the relevant pool of knowledge, scientific fields/academic disciplines have been assigned to relevant industrial fields at the broad level of two-digit ISIC industries using the survey of industrial R&D managers by Levin et al. (1987) to measure the relevance of a discipline to an industry. For example, product innovation activities in drugs (ISIC 24) are linked to research in medicine, biology, chemistry and chemical engineering.

Unfortunately, university research expenditure data disaggregated by scientific disciplines are not available in Austria. But they can be roughly estimated on the basis of two types of data provided by the Austrian Federal Ministry for Science and Research: first, national totals of university research expenditures 1991 disaggregated by broad scientific areas (natural sciences, technical sciences, social sciences, humanities, medicine, agricultural sciences), and, second, data on the number of professional researchers employed in 1991 (that is, university professors, university assistants and contract research assistants) disaggregated by scientific areas and political districts. The best that can be done is to break down the university research expenditure data to the level of scientific disciplines disaggregated by political districts using the following disaggregation procedure:

$$R_{DP} = \frac{R_{AN}}{P_{AN}} P_{DP} \tag{4}$$

where R_{DP} denotes university expenditure in a specific discipline D and in political districts P, R_{AN} national research expenditure in a particular scientific area A, P_{AN} national total of professional researchers in scientific area A, and P_{DP} the number of professional researchers working in university institutes belonging to discipline D and located in political district P.

3   Time-Space Patterns of Knowledge Production in Austria

During the last two decades knowledge production in Austria, measured in terms of patent applications, shows an apparent stability both spatially and by industry. Table 1 shows the sectoral distribution of patent applications in two time periods: 1982-1989 and 1990-1997.

It can be seen from Table 1 that knowledge production concentrates in mechanical areas of manufacturing, especially in machinery. High-technology fields such as electronics, computers or chemistry and pharmaceuticals are significantly less represented. This corresponds to the sectoral structure of manufacturing production (Gassler 1993). However, no apparent specialisation is present at the sectoral level, as indicated by the Herfindahl index (0.30). Neither the total number of patents in manufacturing (about 1,800 per year) nor the ranking of manufacturing sectors (as shown by the high correlation of sectoral shares in the two time periods) have changed meaningfully from the 1980s to the 1990s.

Table 1   Sectoral distribution of Austrian patent applications in the periods 1982-1989 and 1990-1997

	Time period		Percentage change
	1982-1989	1990-1997	
Sectoral share of patents in total patents in manufacturing			
Machinery	26.02	24.52	−5.75
Metal products excluding machines	18.18	19.97	9.87
Instruments	9.48	10.64	12.27
Transportation vehicles	9.23	8.47	−8.29
Chemistry and pharmaceuticals	8.33	7.30	−12.39
Electrical machinery	6.86	6.54	−4.73
Construction	5.53	5.26	−4.88
Stone, clay and glass products	3.73	3.39	−9.10
Paper, printing and publishing	2.53	3.29	30.07
Electronics	2.61	2.78	6.46
Basic metals	2.62	2.52	−3.73
Textiles and clothing	1.87	1.38	−26.49
Computers and office machines	0.77	1.35	75.95
Food, beverages, tobacco	0.83	1.12	34.05
Rubber and plastics	0.94	1.03	9.87
Oil refining	0.29	0.25	−11.77
Wood and furniture	0.18	0.19	5.39
Correlation coefficient	0.99		
Total number of patent applications in manufacturing	15,019	14,251	−5.11
Normalised Herfindahl index of sectoral concentration	0.30	0.29	
Share of vienna in manufacturing total [as percentage]	32.16	34.05	

Source: Austrian Patent Office

Spatial distribution of knowledge production also shows clear stability during the time period of the study. As indicated in Figure 1, there are three larger concentrations of patents and some smaller ones. The three large areas of knowledge production constitute about two-thirds of the total number of Austrian patents. These include the metropolitan area of Vienna (i.e., the city of Vienna and the political districts building the urban fringe) with more than 30 percent of the national knowledge output; the Salzburg and Linz regions with 21 percent, and the Graz region with 8 percent of national knowledge production (see Figure 1).

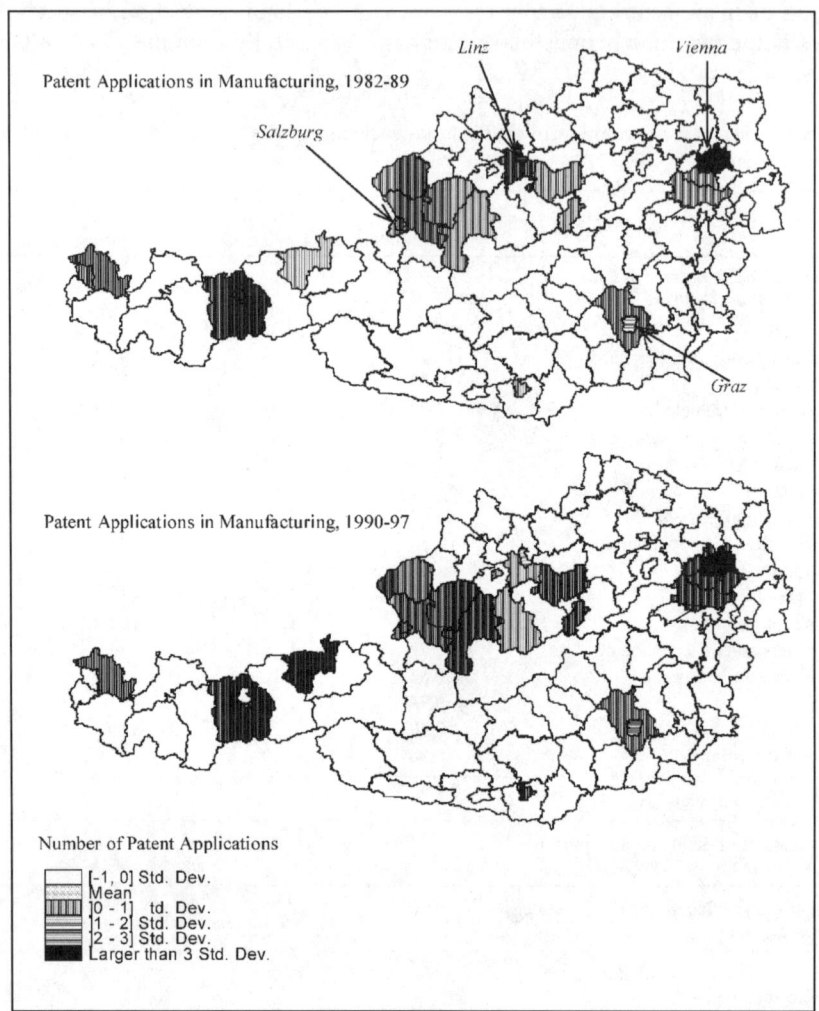

Figure 1   Spatial distribution of Austrian patent applications in manufacturing in the periods 1982-1989 and 1990-1997

Figure 2a provides insights into regional concentration tendencies of Austrian knowledge production for four different manufacturing areas over the period of 1982-1998 measured by means of the normalised Herfindahl Index. There is evidence that electronic industries (including electronics, electrical machinery, computers and office machines), followed by mechanical sectors (such as metal products, machinery, transportation vehicles and instruments) concentrate in a relative small number of political districts, whereas the chemical industries and drugs (chemistry and pharmaceuticals, rubber and plastics, and oil refining) together with traditional sectors (food, beverages and tobacco, construction, stone, clay and glass products, textiles and clothes, paper, printing and publishing, and wood and furniture) tend to spread more widely over the country.

Interestingly, the level of spatial concentration did not change meaningfully during the eighties, whilst the nineties brought a notable decrease in geographical concentration especially in traditional and chemical sectors. This change was induced by a transformation in the spatial structure of Austrian patenting activities. Even though the overall level of knowledge creation was about the same in 1998 (1,637 patents) as it had been in 1982 (1,597 patents), the share of total knowledge output in political districts which had had an above-average level of knowledge creation in the beginning of the period had decreased significantly by the end of the 1990s. The average number of patents diminished from 32 to 24 (a decrease of 25 percent) in political districts with above-average level of knowledge production in 1982, while regions with below-average patent applications at the beginning of the time period expanded their patenting activities from an average of 8 to 14 patents by 1998 (an increase of 88 percent). As a result, the share of those political districts that had above average levels of knowledge creation at the beginning of the period decreased from 72 percent of the national total to 52 percent by the end of the 1990s.

The extent to which political districts with similar levels of knowledge production locate in each other's neighbourhood was measured by the Moran's I statistic for the four manufacturing areas for the period of 1982-1998, and is shown in Figure 2b. A general trend of increasing spatial dependence among neighbouring political districts emerges with no significant variation across industries. However, values of Moran's I remain rather low during the entire period of study and become significant only at the beginning of the 1990s. Some sectoral differences are highlighted in this respect. While for traditional sectors clustering became significant (at the 10 percent level) between 1991 and 1996, this took place for the electronic and mechanical areas during the period of 1995-1998 and 1996-1998, respectively. There was no period of significant spatial clustering for chemical sectors. Overall, the results in Figure 2b show a low level of spatial dependence among neighbouring political districts in Austrian manu-facturing knowledge production.

Figure 2c illustrates the values of the two compatible measures of spatial clustering of knowledge production: the normalised Herfindahl index of geographical concentration and Moran's I statistic of spatial dependence, both calculated at the level of Austrian political districts for manufacturing over the period of 1982 – 1998. The fact that patenting activities did not expand significantly during the period of study together with the opposite trends of the two measures in the 1990s suggests that relocation of knowledge production (as indicated by a decrease in the Herfindahl index) took place from core areas of patenting to their neighbouring political districts (as suggested by the positive trend in Moran's I statistics) resulting in increased spatial concentration of knowledge creation. It is important to note here that the slight increase of clustering in Austrian patenting activities in the period of 1982–1998 does not seem to be the outcome of a dynamic, self-reinforcing process induced by local environments with many knowledge externalities leading to expanding clusters of knowledge production and

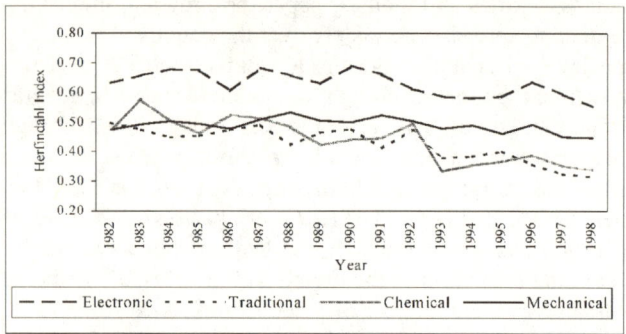

Figure 2a Geographical concentration of patents for four manufacturing areas, measured by the normalised Herfindahl index (1982-1998)

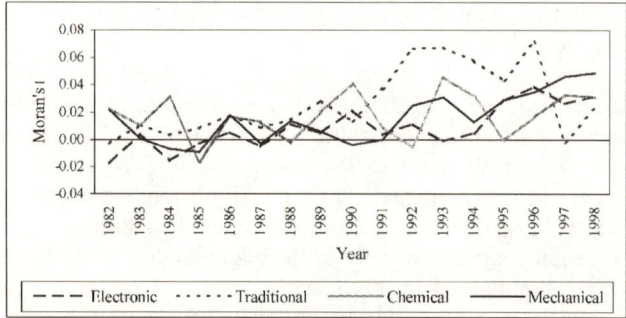

Figure 2b Spatial association of patents across political districts for four manufacturing areas, measured by Moran's *I* statistics (1982-1998)

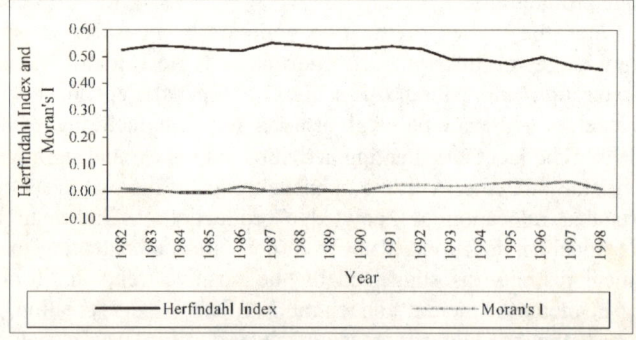

Figure 2c Geographic concentration and spatial association across political districts of patents in manufacturing in Austria (1982-1998)

an overall growth in knowledge output. Instead, it is characterised by a spatial shift of knowledge production to neighbouring peripheral areas, while the overall level of knowledge output stays largely unchanged.

It is important to note here that the slight increase of clustering in Austrian patenting activities in the period of 1982–1998 does not seem to be the outcome of a dynamic, self-reinforcing process induced by local environments with many knowledge externalities leading to expanding clusters of knowledge production and an overall growth in knowledge output. Instead, it is characterised by a spatial shift of knowledge production to neighbouring peripheral areas, while the overall level of knowledge output stays largely unchanged.

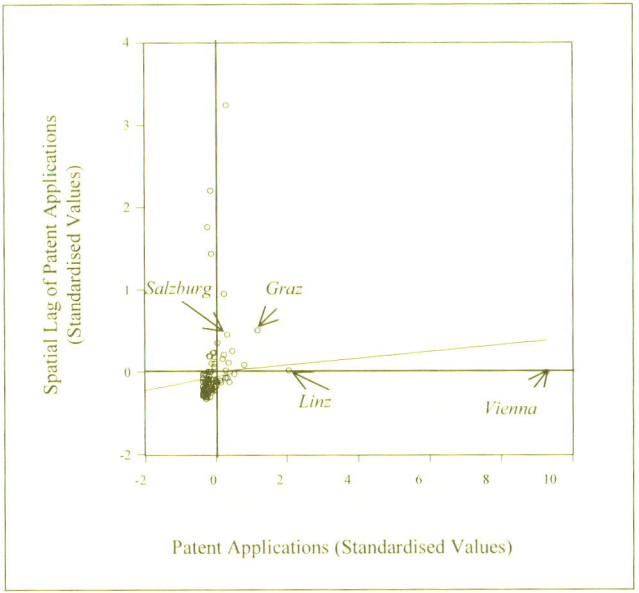

Figure 3   Moran scatterplot: Austrian patent applications in manufacturing (1998)

The *Moran scatterplot* of Austrian patents in 1998 in Figure 3 shows spatial patterns of Austrian knowledge production at the end of the study period. The units of observational are Austrian political districts. The horizontal axis represents standardised values of patent counts while on the vertical axis average values of the same variable in neighbouring political districts are given (i.e., a row-standardised simple contiguity matrix is used for calculations). The positive slope of the regression line reflects a positive value of Moran's I indicating an overall tendency of positive spatial association among neighbouring political districts. This tendency is predominantly supported by spatial clustering of political districts where lower than average level of knowledge creation takes place (as indicated by the high concentration of observations in the lower left quadrant of the scatterplot). Leverage points in the upper right quadrant (i.e., political districts with above average patenting activity neighbouring similar regions) include Salzburg, Linz and Graz.

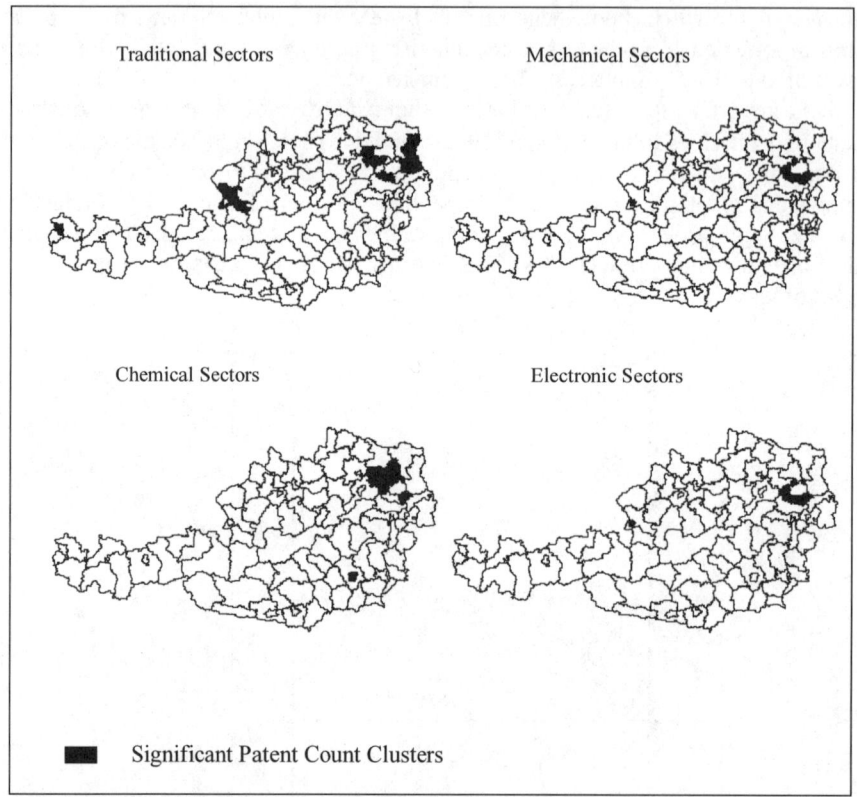

Traditional Sectors   Mechanical Sectors

Chemical Sectors   Electronic Sectors

■■■   Significant Patent Count Clusters

Figure 4   Clusters of high values of patent counts for four manufacturing areas, measured
by significant values of the local Moran statistics (1998)

It is very clear from Figure 3 that Vienna is an outlier in Austrian knowledge
production. The standardised value of the number of patents in Vienna is nine
times higher than the respective Austrian average. On the other hand, it is also
demonstrated that Vienna is surrounded by political districts with levels of
patenting activities around the average (i.e., the mean value of patents in its
neighbourhood equals the national average).

Significant clusters of patenting activity in four manufacturing areas in 1998
are shown in Figure 4. Significance at $p < 0.05$ is based on 1,000 random
permutations. A row-standardised simple contiguity matrix has been used for
calculations. Dark areas stand for core political districts of spatial clusters. The
largest clusters are formed in traditional sectors whereas mechanical and
electronic concentrations are relatively small. The Vienna metropolitan area is a
significant cluster in all areas of manufacturing. Other clusters are formed around
Salzburg (traditional, mechanical and electronic sectors), Graz (chemical sectors)
and Dornbirn at the western border of the country (traditional sectors).

4   Local Inputs to Innovation – An Assessment of Their Relative Significance in Knowledge Production

As emphasised in the literature on innovation systems, production of new technological knowledge is not simply the outcome of the independent efforts of firms to innovate, but is also influenced by knowledge interactions with various actors in the system including other firms, and private and public research institutions. However, knowledge flows are very difficult (if not impossible) to trace empirically. Different methods have been proposed in the literature to measure knowledge flows at least partially such as patent citation analysis (Jaffe et al. 1993), analysis of patterns of co-patenting or co-publications (Hicks and Katz 1996) and counts of industry technology alliances (Haagedoorn 1994).

In the previous section we observed that there has been a slightly increasing, but still a relatively modest level of geographical clustering of knowledge production in Austria. Since no systematically collected data on knowledge interactions are available at the level of the regions, in this section we have applied an indirect approach to assess the significance of local inputs to knowledge production. A positive association in the spatial distribution of patenting and local knowledge inputs is taken to be an indication of knowledge spillovers existing in the production of economically useful new technological knowledge. Industrial R&D and university research are considered as potentially providing direct inputs to knowledge production, whereas manufacturing employment is included in the analysis as a proxy for unspecified agglomeration effects. Analysis is based on data aggregated at the level of Austrian political districts. In order to account for the time necessary to come up with patentable inventions, following the industrial experience reported for example in Edwards and Gordon (1984), a two-year time lag is applied between knowledge inputs (1991) and knowledge output (1993).

Table 2 provides a general profile of sectoral distribution of the three proxy variables of inputs to knowledge production: R&D in manufacturing and university research expenditures as well as the auxiliary variable of manufacturing employment. The three variables evidently follow different patterns of sectoral specialisation. Whereas R&D in manufacturing concentrates in electronics, university research focuses mainly on chemistry and pharmaceuticals, and instruments. On the other hand, about fourty percent of manufacturing employment is in the machinery, food and wood sectors. However, the overall sectoral concentration is not very strong, especially in manufacturing employment as indicated by the corresponding Herfindahl index. Low values of correlation coefficients with patent counts in manufacturing suggest that sectoral distribution of knowledge production at the country level only vaguely follows the respective patterns of R&D, university research and employment.

Figure 5 shows that, though by and large the spatial distribution of patent counts follows the geographical patterns of industrial and university R&D as well as manufacturing employment, there are notable differences in pattern matching. A deeper understanding of the geographical patterns of Austrian knowledge production may be gained by calculating correlation coefficients between patent counts and each of the input measures (including the auxiliary variable of employment) at the level of political districts and for four manufacturing areas[1], in order to account for the supposedly different characteristics of the innovation system of the metropolitan area of Vienna (the definite positive outlier in Austrian knowledge production) and the three major cities supporting the overall positive clustering tendency of patent counts (i.e., Salzburg, Linz and Graz).

Table 2   Sectoral distribution of R&D in manufacturing, university research and manufacturing employment (1991) (ranking follows patent orders in 1990-1997 in Table 1)

Manufacturing sectors [a]	R&D expenditure in manufacturing	University research expenditures	Manufacturing employment
Machinery	11.78	11.22	12.64
Metal products excluding machines	3.11	9.07	10.09
Instruments	0.73	59.49	3.80
Transportation vehicles	7.05	21.58	4.62
Chemistry and pharmaceuticals	15.22	62.41	4.13
Electrical machinery	7.67	11.81	4.18
Stone, clay and glass products	5.04	4.06	6.06
Paper, printing and publishing	2.00	na	7.19
Electronics	29.68	11.81	2.22
Basic metals	4.28	9.07	5.14
Textiles and clothing	1.72	na	9.43
Computers and office machines	1.98	25.27	0.14
Food, beverages, tobacco	2.20	1.55	12.45
Rubber and plastics	5.86	9.23	4.13
Oil refining	1.37	9.23	0.37
Wood and furniture	0.32	0.80	13.40
Manufacturing total	16.25[b]	6.41[b]	0.72[c]
Normalised Herfindahl index of sectoral concentration	0.31	na	0.15
Correlation with the sectoral share of patents in 1990-1998	0.15	na	0.34

[a] denotes column percentage (for R&D expenditures in manufacturing and employment) and percentages of total university R&D expenditures (for university research expenditures). Given that certain university institutes are allocated to more than one manufacturing sector, sum of percentages is not 100 in the third column. [b] is in terms of 10^9 ATS. [c] is in terms of 10^6 persons.

---

[1]   Traditional sectors include food, beverages and tobacco [ISIC 15-16], construction [ISIC 45], stone, clay and glass [ISIC 25], textiles and clothing [ISIC 17 and 18], paper, printing and publishing [ISIC 21 - 22] and wood and furniture [ISIC 20 and 36]. The mechanical sectors include basic metals [ISIC 27], instruments [ISIC 33], transportation vehicles [ISIC 34 - 35], machinery [ISIC 29] and metal products [ISIC 28]. The chemical sectors consist of rubber and plastics [ISIC 25], chemistry and pharmaceuticals [ISIC 24] and oil refining [ISIC 23], whereas the electronic sectors include electronics [ISIC 32], electrical machinery [ISIC 31] and computers and office machines [ISIC 30].

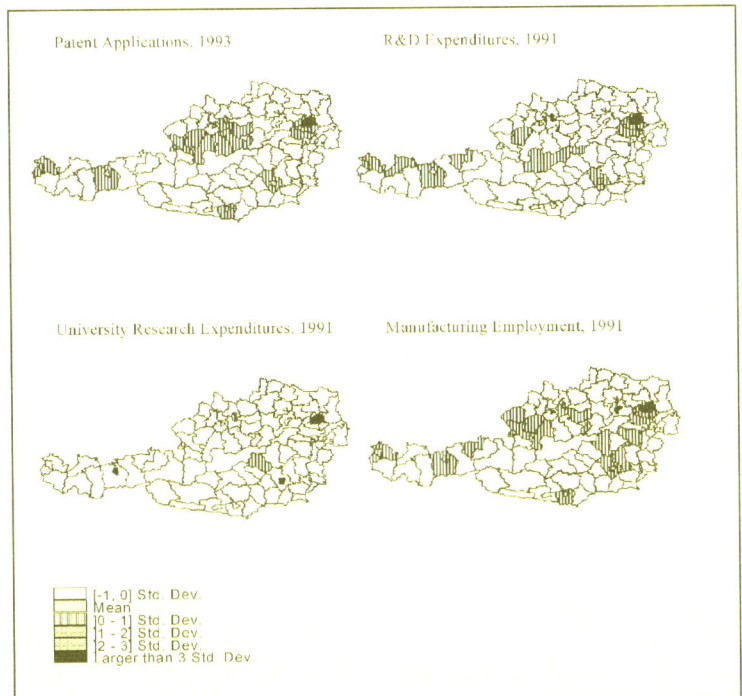

Figure 5   Spatial distribution of patent applications, private R&D expenditures, university research expenditures and manufacturing employment in Austria

Figure 6 shows correlation coefficient values for three different sets of observations: the whole sample, political districts excluding Vienna and political districts excluding Salzburg, Linz, Graz and Vienna. The following three major observations can be derived from this figure. First, the four manufacturing areas exhibit dissimilar correlation patterns. Considering only those coefficients calculated for the whole sample, patent counts in electronic sectors are highly correlated with all the three measures of local knowledge inputs, while knowledge production in chemicals is more related to local employment and R&D. On the other hand, in mechanical and traditional sectors the highest correlations are observed with employment and university research. Second, the data shown in the figure suggests that the outlier position of Vienna in knowledge production might well be the result of its comparatively strong reliance on local knowledge inputs.

After excluding Vienna from the sample, correlation coefficients decrease significantly, especially the R&D measures. The smallest falls are observed in correlations with local employment, with the exception of the electronic sectors. Third, regarding the degree to which knowledge production in the three major Austrian cities exhibit distinct characteristics relative to the rest of the sample (excluding Vienna), dissimilar patterns are observed for the research variables, but not for employment.

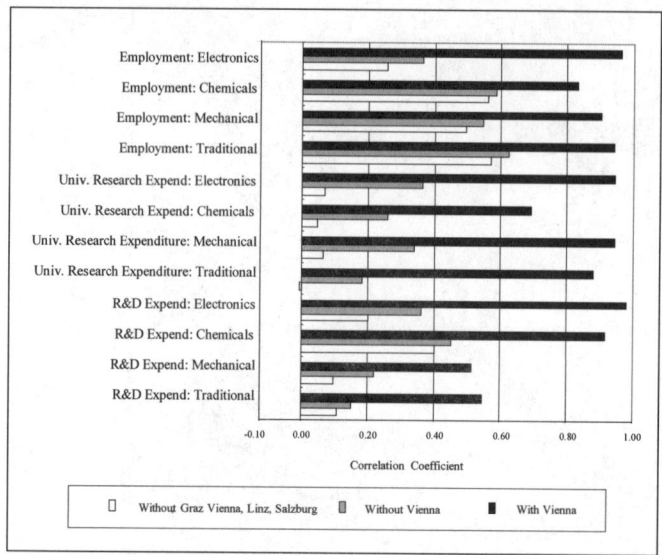

Figure 6   Correlation between patents in 1993 and selected measures of potential local inputs to innovation in 1991 at the level of Austrian political districts

It is important to note that regional knowledge output increases faster than any of its local inputs. This might be taken as a sign of the existence of regionally mediated knowledge flows. In Figure 7 we can see scatterplot diagrams of patents and R&D in manufacturing, university research and manufacturing employment. Data are arranged in increasing order of the variables on the horizontal axes.

Additionally, in order to give an indication of the direction and size of the change in patents in manufacturing, we have also estimated the curves of nearest neighbour fit [Loess fit]. For each data point in the sample, a locally weighted polynomial regression has been estimated. This is a local regression, since we have used only the subset of observations which lie in the neighbourhood of each point to fit the regression model (Cleveland 1994). In case of increasing returns in knowledge production, Loess fit curves show an exponential growth in patents.

The only variable for which increasing returns dominate the entire sample is manufacturing employment. This shows that the higher the concentration of production in an area, the higher the probability of knowledge related linkages arising among firms, resulting in a higher than proportional increase in knowledge production. However, this relationship can not be observed for R&D in manufacturing and university research linkages throughout the whole sample. Some degree of potential research spillover effects might be present in larger cities, and they seem to have a definite role in Vienna (the highest point in each scatterplot). However, Figure 8 indicates no signs of significant interregional linkages.

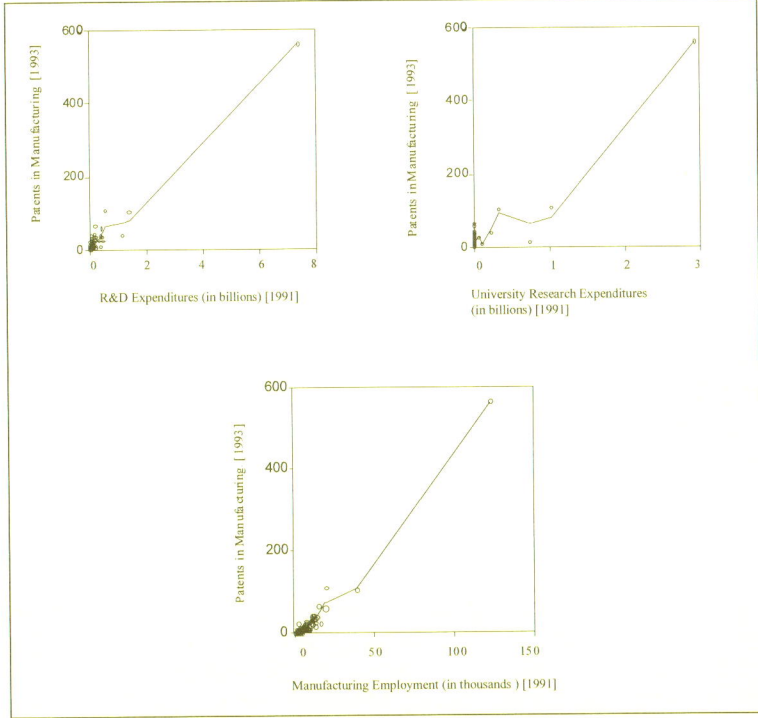

Figure 7   Scatterplots with curves of nearest neighbour fit [Loess fit] for patents
in manufacturing related to R&D in manufacturing, university research
and manufacturing employment in Austria

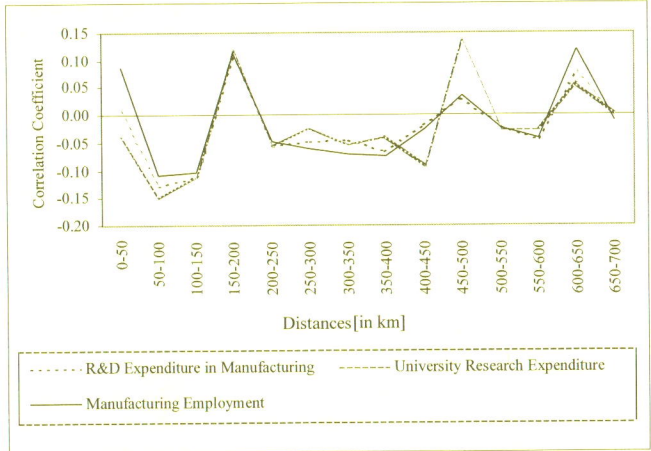

Figure 8   Cross-regional correlation patterns between patent applications (1993) and knowledge inputs (1991) in increasing
distances from the patenting political district

5 Conclusions

In recent years, the role of space in general and of spatial externalities in particular has gained an increasingly prominent position in mainstream economics, partly stimulated by the visibility of Krugman's work on the New Economic Geography (Krugman 1991). Of course, the importance of space is not new to geographers and regional scientists. Based on descriptive and exploratory techniques (Moran's *I* test for spatial autocorrelation and the Moran scatterplot) in this chapter we have made an initial attempt to analyse the effect of space in the creation of knowledge. Clusters of the output of the knowledge creation process (measured in terms of patent counts) are compared with spatial concentration patterns of three input measures of local knowledge production: R&D in manufacturing, university research activities and manufacturing employment.

Empirical evidence shows that knowledge production in Austria tends to focus largely in mechanical areas of manufacturing rather than in high-tech fields such as the electronic sectors. It is interesting to note that this pattern has changed little during the past two decades. We have been able to identify only a weakly growing trend of clustering. However, this does not appear to be so much the outcome of a dynamic process generated by intensive knowledge flows at the local level, as the consequence of a spatial shift in knowledge production. There is no doubt that Vienna with its strong presence of high quality research organisations and R&D in manufacturing dominates the knowledge creation process. Some smaller clustering tendencies were discovered around Salzburg, Linz and Graz.

Geographic stability of knowledge generation characterised by weakly expanding clusters may well be the outcome of relatively undeveloped linkages among the major actors of the Austrian innovation system as suggested by the limited role of local knowledge flows in most parts of the country. Cluster generating increasing returns appear to result largely from between-firm knowledge diffusion rather than from knowledge spillover effects.

As in the case of any exploratory data analysis, the above findings need to be treated with caution and should be viewed only as an initial pre-modelling stage in the endeavour. Future research activities will be devoted to shedding further light on the issue of local university knowledge transfer by transforming the analytical Griliches-Jaffe knowledge production approach (see, Griliches 1979; Jaffe 1989), modelling knowledge spillovers in form of a spatially discounted external stock of knowledge, and employing spatial econometric tools for model specification and estimation (see Fischer and Varga 2001).

Acknowledgements: The authors gratefully acknowledge the grant no. 7994, provided by the Jubiläumsfonds of the Austrian National Bank, and the support received from the Department of Economic Geography & Geoinformatics at the Vienna University of Economics and Business Administration, and the Austrian Research Centers Seibersdorf. They also wish to express their thanks to Christian Rammer, Doris Schartinger, Norbert Böck [Austrian Research Centers Seibersdorf], Werner Hackl [Austrian Chamber of Commerce, Vienna] and Karl Messman [Austrian Central Statistical Office, Vienna] for assisting in various phases of data collection.

References

Anselin L. (1997): The Moran scatterplot as an ESDA tool to assess local instability in spatial association. In: Fischer M., Scholten H. and Unwin D. (eds.) *Spatial Analytical Perspectives on GIS in Environmental and Socio-Economic Sciences*, Taylor & Francis, London, pp. 111-125

Anselin L. (1995): Local indicators of spatial association – LISA, *Geographical Analysis* 27 (2), 93-115

Archibugi D. (1992): Patenting as an indicator of technological innovation: A review, *Science and Public Policy* 19 (6), 357-368

Archibugi D. and Pianta M. (1992): *The Technological Specialisation of Advanced Countries. A Report to the EEC on International Science and Technology Activities*, Kluwer, Dordrecht, Boston

Basberg B. (1987): Patents and the measurement of technological change: A survey of the literature, *Research Policy* 16 (2-4), 131-141

Braczyk H.-J., Cooke P. and Heidenreich M. (1998): *Regional Innovation Systems: The Role of Governance in a Globalised World*, UCL Press, London

Cleveland W. (1994): *The Elements of Graphic Data*, Hobart Press, Summit

Cooke P., Uranga M. and Etxebarri G. (1997): Regional innovation systems: Institutional and organisational dimensions, *Research Policy* 26 (4/5), 475-491

Edwards K. and Gordon T. (1984): *Characterization of Innovations Introduced on the U.S. Market in 1982*, The Futures Group, U.S. Small Business Administration, Washington D.C.

Fischer M.M. (2001a): Innovation, knowledge creation and systems of innovation, *The Annals of Regional Science* 35 (2), 199-216

Fischer M.M. (2001b): Spatial analysis in geography. In: Smelser N.J. and Baltes P.B. (eds.) *International Encyclopedia of the Social and Behavioral Sciences*, Vol. 22, Elsevier, Oxford, pp. 14752-14758

Fischer M.M. (1998): Spatial analysis: Retrospect and prospect. In: Longley P., Goodchild M.F., Maguire D.J. and Rhind D.W. (eds.) *Geographical Information Systems: Principles, Technical Issues, Management Issues and Applications*, Wiley, New York, pp. 283-292

Fischer M.M. and Varga A. (2001): Geographic knowledge spillovers and university research: Some evidence from Austria. In: Institute for Geography and Regional Research, University of Vienna (ed.), *Geographischer Jahresbericht aus Österreich* 58, pp. 37-47

Fischer M.M., Fröhlich J. and Gassler H. (1994): An exploration into the determinants of patent activities: Some empirical evidence for Austria, *Regional Studies* 28 (1), 1-12

Freeman C. (1987): *Technology and Economic Performance: Lessons from Japan*, Pinter, London

Gassler H. (1993): Regionale Disparitäten der betrieblichen Inventionsaktivitäten in Österreich. Eine empirische Analyse unter Verwendung von Patentdaten, *Klagenfurter Geographische Schriften* 11, 173-186

Griliches Z. (1990): Patent statistics as economic indicators: A survey, *The Journal of Economic Literature* 28 (4), 1661-1707

Griliches Z. (1979): Issues in assessing the contribution of research and development to productivity growth, *Bell Journal of Economics* 10, 92-116

Hagedoorn J. (1994): Technological partnering in strategic alliances, Paper prepared for the Austrian Conference on R&D Co-Operation, Vienna

Hicks D. and Katz S. (1996): Systemic bibliometric indicators for the knowledge-based economy, Paper presented at the OECD Workshop on New Indicators for the Knowledge-Based Economy, Paris, 19-21 June, 1996

Jaffe A.B. (1989): Real effects of academic research, *American Economic Review* 79, 957-970

Jaffe A.B., Trajtenberg M. and Henderson R. (1993): Geographic localisation of knowledge spillovers as evidenced by patent citations, *Quarterly Journal of Economics* 108 (3), 577-598

Krugman P. (1991): Increasing returns and economic geography, *Journal of Political Economy* 99 (3), 483-499

Levin R.C., Klevorick A.K., Nelson R.R. and Winter S.G. (1987): Appropriating the returns from industrial research and development, *Brookings Papers on Economic Activity* 1987 (3), 783-820

Pavitt K. (1988): Uses and abuses of patent statistics. In: Raan A.F.J. van (ed.) *Handbook of Quantitative Studies of Science and Technology*, North-Holland, Amsterdam, pp. 509-535

Storper M. (1997): *The Regional World. Territorial Development in a Global World*, The Guilford Press, New York, London

Verspagen B., van Moergastel T. and Slabbers M. (1994): MERIT Concordance Table: IPC-ISIC (rev.2), MERIT Research Memorandum 2/94-004, Maastricht Economic Research Institute on Innovation and Technology, University of Limburg

Appendix A: Patent Applications, R&D Expenditures, University Research Expenditures and Employment in Manufacturing for 99 Austrian Political Districts

Political district	Patent applications (1993)	Industrial R&D expenditures (in 10^3 ATS, 1991)	University research expenditures (in 10^6 ATS, 1991)	Employment (1991)
Eisenstadt (Stadt)	6	0	0	596
Rust (Stadt)	0	0	0	41
Eisenstadt-Umgebung	4	32,344	0	1,776
Güssing	0	1,000	0	1,023
Jennersdorf	1	0	0	1,427
Mattersburg	4	10,548	0	3,461
Neusiedl am See	5	14,771	0	1,731
Oberpullendorf	1	4,390	0	2,555
Oberwart	0	3,978	0	5,096
Klagenfurt (Stadt)	27	13,527	36	7,113
Villach (Stadt)	11	25,919	0	5,647
Hermagor	2	160	0	1,045
Klagenfurt Land	22	0	0	2,251
Sankt Veit an der Glan	5	8,160	0	5,162
Spittal an der Drau	5	90,711	0	4,655
Villach Land	8	52,886	0	3,687
Völkermarkt	1	3,200	0	3,236
Wolfsberg	5	18,586	0	4,497
Feldkirchen	2	1,439	0	1,702
Krems (Stadt)	5	52,877	0	4,057
Sankt Pölten (Stadt)	7	33,383	0	8,333
Waidhofen (Stadt)	7	4,595	0	1,606
Wiener Neustadt (Stadt)	5	36,376	0	5,143
Amstetten	36	107,121	0	12,255
Baden	38	348,885	0	13,350
Bruck an der Leitha	2	34,450	0	2,343
Gänserndorf	13	7,225	0	4,711
Gmünd	9	0	0	5,514
Hollabrunn	1	770	0	1,743
Horn	7	1,456	0	2,279
Korneuburg	16	26,586	0	5,579
Krems (Land)	2	0	0	1,823
Lilienfeld	1	3,521	0	3,253
Melk	9	9,790	0	4,714
Mistelbach	8	0	0	3,697
Mödling	32	196,105	0	10,616
Neunkirchen	13	67,802	0	8,637
Sankt Pölten (Land)	14	51,303	0	7,303
Scheibbs	2	3,600	0	2,847
Tulln	4	28,057	0	3,445
Waidhofen an der Thaya	1	11,930	0	3,168
Wiener Neustadt (Land)	7	7,618	0	5,515
Wien-Umgebung	23	305,350	0	10,303
Zwettl	4	0	0	2,233
Linz (Stadt)	101	1,375,777	218	39,068
Steyr (Stadt)	39	1,124,624	0	11,399
Wels (Stadt)	28	35,720	0	9,744
Braunau am Inn	14	158,617	0	12,958
Eferding	5	3,772	0	2,725
Freistadt	1	420	0	2,571
Gmunden	42	133,864	0	11,832

Appendix A (*ctd.*)

Political district	Patent applications (1993)	Industrial R&D expenditures (in 10^3 ATS, 1991)	University research expenditures (in 10^6 ATS, 1991)	Employment (1991)
Grieskirchen	14	51,170	0	5,883
Kirchdorf an der Krems	23	17,706	0	7,065
Linz-Land	21	102,877	0	16,499
Perg	13	23,580	0	4,894
Ried im Innkreis	7	50,189	0	6,108
Rohrbach	4	3,650	0	3,817
Schärding	8	33,760	0	4,239
Steyr-Land	12	9,314	0	3,317
Urfahr-Umgebung	10	0	0	2,658
Vöcklabruck	56	386,655	0	19,110
Wels-Land	9	79,982	0	7,511
Salzburg (Stadt)	37	41,309	137	10,594
Hallein	12	123,539	0	6,642
Salzburg-Umgebung	31	22,640	0	10,490
Sankt Johann im Pongau	14	21,155	0	5,200
Tamsweg	1	0	0	1,044
Zell am See	7	32,316	0	4,575
Graz (Stadt)	105	519,747	1,288	19,544
Bruck an der Mur	7	99,697	0	9,246
Deutschlandsberg	9	114,536	0	5,595
Feldbach	3	6,705	0	4,050
Fürstenfeld	2	12,416	0	2,308
Graz-Umgebung	25	461,144	0	9,425
Hartberg	4	10,400	0	4,929
Judenburg	14	79,326	0	6,633
Knittelfeld	3	19,529	0	3,805
Leibnitz	4	3,017	0	5,377
Leoben	9	48,238	176	6,755
Liezen	7	191,806	0	6,040
Mürzzuschlag	6	26,212	0	6,336
Murau	4	0	0	1,837
Radkersburg	0	383	0	1,249
Voitsberg	13	40,615	0	4,010
Weiz	9	142,596	0	7,566
Innsbruck (Stadt)	15	5,692	907	5,637
Imst	5	14,050	0	2,352
Innsbruck (Land)	35	422,458	0	13,247
Kitzbühel	10	22,031	0	3,233
Kufstein	10	356,486	0	9,382
Landeck	0	0	0	1,776
Lienz	5	9,147	0	4,043
Reutte	5	183,676	0	2,722
Schwaz	18	102,295	0	7,303
Bludenz	5	24,674	0	7,075
Bregenz	65	180,774	0	14,763
Dornbirn	23	191,232	0	13,117
Feldkirch	20	134,127	0	10,918
Vienna	541	7,374,721	3,652	122,960

Sources: Patent applications data come from the Austrian Patent Office; industrial R&D data from the Austrian Chamber of Commerce; university research data from the Austrian Federal Ministry for Science and Research; employment data from the Austrian Central Statistical Office.

10 Spatial Knowledge Spillovers and University Research: Evidence from Austria

with *A. Varga*

This chapter provides some evidence on the importance of geographically mediated knowledge spillovers from university research activities to regional knowledge production in high-technology industries in Austria. Spillovers occur because knowledge created by universities has some of the characteristics of public goods, and creates value for firms and other organisations. The chapter lies in the research tradition that finds thinking in terms of a production function of knowledge useful and looks for patents as a proxy of the output of this process, while university research and corporate R&D investment represent the input side. It refines the classical regional knowledge production function by introducing a more explicit measure to capture the pool of relevant spatial academic knowledge spillovers. A spatial econometric approach is used to test for the presence of spatial effects and – when needed – to implement models that include them explicitly. The empirical results confirm the presence of geographically mediated university spillovers that transcend the spatial scale of political districts. They, moreover, demonstrate that such spillovers follow a clear distance decay pattern.

1 Introduction

Innovation activities involve the use, application and transformation of scientific and technical knowledge in the solution of practical problems. Much of the essential knowledge in this process is specialised and resides in tacit form within experienced researchers and engineers. Tacitness refers – as Dosi (1988, p. 1126) suggested on the basis of earlier insights by Polanyi (1967) – to 'those elements of knowledge, that persons have, which are ill-defined, uncodified, and which they themselves can not articulate, and which differ from person to person, but which to some degree be shared by collaborators who have a common experience'. This kind of knowledge has to be carefully distinguished from information in the usual sense that is factual, while knowledge is characteristically complex and aims to discover the *why* (procedural knowledge) and *how* (skills and competences).

Knowledge has some of the characteristics of public goods. It is widely considered to be a partially excludable and non-rivalrous good (see Romer 1990). Non-rivalry implies that a novel piece of knowledge can be utilised many times and in many different circumstances without reducing its value. Knowledge is only imperfectly excludable and, thus, subject to spillovers. One might view knowledge spillovers as leaks, but in reality they are the sine qua-non condition for the development of knowledge and economic growth (OECD 1992, Romer 1990). Following Cohen and Levinthal (1989) we define knowledge spillovers to include 'any original, valuable knowledge generated in the research process which becomes publicly accessible, whether it be knowledge fully characterising an innovation, or knowledge of a more intermediate nature'.

In this paper we will concentrate on knowledge spillovers[1] that originate from university research. There are numerous channels through which knowledge might spread to firms. It may seep into the public domain in publications or public presentations of various types (university seminars, academic conferences etc.). It may travel with graduates who take a job at a firm or start their own. It may also be uncovered through reverse engineering and other purposive search processes. The extent to which knowledge flows through these different channels depends upon the capability of the recipient (especially, his/her absorptive capacity), the nature of the knowledge itself (for example, whether it is tacit or codified), and other factors that bring academic and industry sector researchers together (Geroski 1995). If knowledge is essentially tacit, then it can not be transferred by ways other than personal interaction, and geographical distance matters. Thus, the creation of knowledge is a process that is essentially localised.

Since knowledge spillovers are not directly observable, systematic evidence on the extent and importance of such spillovers is difficult to come by. In recent years various attempts have started to document the effect of academic knowledge spillovers on corporate R&D in manufacturing industry, almost exclusively in a US American context. Research by Nelson (1986); Mansfield (1991, 1995); Jaffe (1989); Adams (1990, 1993); Ács et al. (1992, 1994), and others has found that university research has substantial effects on technological change in important segments of the economy[2]. Using state-level patent and innovation data, respectively, Jaffe (1989), Ács et al. (1992) and others have added an important spatial dimension to the discussion by illustrating that the effects not only differ by industries, but also increase with geographic proximity.

These and many other studies that followed[3] did find a strong and positive relationship between patenting or innovative activity, and university research and corporate R&D at the state level in the US. The situation, however, is different in terms of the significance of *local* geographic spillover effects. Overall considered the evidence is non-existent, weak or mixed, and only pertaining to a few individual sectors (see, for example, Anselin et al. 2000). This lack of evidence

---

[1]   More precisely on 'pure' knowledge spillovers in contrast to rent spillovers that are closely linked to knowledge embodied in traded capital or intermediated goods.

[2]   Most have used the production function approach inspired by Griliches (1979) and Jaffe (1989), some (see, for example, Bernstein and Nadiri 1988) the cost function approach to estimate the effects of spillovers. The disadvantage of the latter approach is the required use of prices.

[3]   For a survey of the literature see Karlsson and Manduchi (2001).

contradicts the strong findings in micro-level studies (see, for example, Mansfield 1995; Jaffe et al. 1993).

The objective of this paper is to shed some further light on the issue in an Austrian context. The study lies in the research tradition inspired by Griliches (1979) and Jaffe (1989), but departs from previous research in two major respects. First, it is based on a much finer, and thus, more appropriate spatial scale than most previous studies to capture interactions between universities and high-technology based firms. Second, we specify the relevant potential of spillovers in form of spatially discounted pools of knowledge. The specification makes use of accessibility measures derived from established principles in spatial interactions theory[4]. A spatial econometric approach is implemented both by testing for the presence of spatial effects and – when necessary – by implementing models that incorporate them explicitly. In the remainder of the paper we first introduce the conceptual framework in Section 2. Next we briefly describe the variables and the data sets (Section 3), then outline subsequently some methodological issues in specifying and estimating the model (Section 4) and finally present the results obtained (Section 5). The paper concludes with a brief evaluation of the results associated with some hints for future research activities.

2 The Knowledge Production Function

We adopt the view that finds thinking in terms of a production function of knowledge congenial and useful, and looks for patents or innovations to serve as a proxy of the output of this process, while university research and commercial R&D represent the input side. Less 'neoclassical' oriented economists might deny the usefulness of this view or the simplifications on which this view is based. But we believe that the importance and extent of academic knowledge spillovers can be best discussed in the context of an empirically useful regional variant of the knowledge production function.

The basic model relates the output of the process, the increment of economically valuable technological knowledge (say, K), in region i ($i = 1, ..., N$) to research and development inputs. Regional knowledge production may be seen to depend on two major sources[5]: University research, say U, and commercial research and development, say R, located in region i. Inventive inputs have

---

[4] See Frost and Spence (1995) for a recent review of spatial accessibility measures.

[5] The main institutions created by Western Society to meet the purpose to generate fundamental, general and public knowledge have been its universities and learned societies. Fundamental research of the quality and on the scale comparable to these institutions calls for high thresholds of R&D investment and a corporate research environment conducive to developing and discussing ideas freely with other research workers. Knowledge development within firms also raises proprietary issues. Thus, some sort of division of labour has been developed between university research on the one side and industry R&D on the other (see OECD 1992).

generally been treated as measured by the resources invested in them, most often research and development expenditures. The underlying assumption in general (see, for example, Anselin et al. 1997 and many others) is to assume that research and development expenditures will lead to immediate inventive results. Because the production of useful knowledge takes time, we depart from this common practice and assume a time lag between the investment and the yield of results. Thus, our basic regional knowledge production function is given in general form as

$$K_{i,t} = f(U_{i,t-q}, R_{i,t-q}) \qquad \text{for } i = 1, ..., N \tag{1}$$

where the subscripts i and t refer to region i and time t, respectively. q denotes the time shape of the lag between research investment and invention results. $U_{i,t-q}$ and $R_{i,t-q}$, represent university research and industry R&D investments, respectively. We may call this equation – more precisely f – the *classical regional knowledge production function*.

Of course, this formulation is rather simplistic and is based on several simplifying assumptions, either explicit or implicit. For example, implicit is the assumption that the production of knowledge of a particular firm or industry not only depends on its own research efforts, but also on outside efforts or – more generally – on the knowledge pool available within the region. It is assumed that knowledge generated in universities spills over to the generation of economically valuable technological knowledge by firms. Moreover, generally the assumption is made that the variable U represents the local pool of potential university spillovers. Knowledge tacitness is the reason for the local dimension of spillovers.

The model is comparative-static in nature and abstracts from some important dynamic issues. In particular, there are long, variable, and uncertain lags in the interval between the start of a research activity and generating useful knowledge. The implicit assumption of a stable relationship between the input of the production process (U and R) and its output (in terms of K) may be defended on the perception that science progresses in general by a sequence of marginal improvements rather than through a series of discrete, essentially sporadic breakthroughs (see, for example, Kamien and Schwartz 1982; Rosenberg 1976). Assumptions about the properties of f – such as diminishing returns to research expenditures or economies of scale and economies of scope – imply restrictions on the relationship between (U, R) and K.

The increment to useful knowledge arising from R&D and university research is likely to depend upon a number of further factors including a host of variables related to the institutional and management environment within which the resources are deployed. We may broaden model (1) by including these additional influences represented by a vector of variables, Z_i, that reflects these additional influences. Thus

$$K_{i,t} = f(U_{i,t-q}, R_{i,t-q}, Z_{i,t-q}) \qquad \text{for } i = 1, ..., N. \tag{2}$$

The problem of modelling regional knowledge production is much more complicated when we realise that different amounts of knowledge from different regions may spill-in. There are different approaches to the construction of spillover stocks or pools. We utilise the approach where every possible pair of regions is treated separately, and the relevant stock of non-local spillovers for the receiving region is constructed specifically for it, using its distance from the $N - 1$ spilling regions as a weight. There is a wide choice of possible weights. We use a spatial accessibility measure to induce a distance metric[6].

To simplify notation, let us denote

$$U'_{t-q} = (U_{1,t-q}, ..., U_{N,t-q}) \tag{3}$$

$$R'_{t-q} = (R_{1,t-q}, ..., R_{N,t-q}) \tag{4}$$

and

$$D_{i\bullet} = (d_{i,1}^{-\gamma}, ..., d_{i,i-1}^{-\gamma}, 0, d_{i,i+1}^{-\gamma}, ..., d_{i,N}^{-\gamma}) \qquad \text{for } i = 1, ..., N \tag{5}$$

where d_{ij} represents the average geographic distance from the spilling region j ($j \neq i$) to the receiving region i. $\gamma > 0$ is a distance decay parameter. Then we can define the spatially discounted pool of non-local university spillovers as

$$S_{i,t-q}^{U} = D_{i\bullet} U_{t-q} \qquad \text{for } i = 1, ..., N \tag{6}$$

and the spatially discounted pool of non-local industry R&D spillovers as

$$S_{i,t-q}^{R} = D_{i\bullet} R_{t-q} \qquad \text{for } i = 1, ..., N. \tag{7}$$

This yields the following regional knowledge production function in general form

$$K_{i,t} = f(U_{i,t-q}, S_{i,t-q}^{U}, R_{i,t-q}, S_{i,t-q}^{R}, Z_{i,t}) \qquad \text{for } i = 1, ..., N \tag{8}$$

that will enable us to capture intra- and interregional knowledge spillovers of two types, those originating from university research and those from industrial R&D.

In order to implement model (8) we need to specify the functional form of f. For the purpose of this study we have taken the Cobb-Douglas version which can be written in logarithmic form as

$$\begin{aligned} \log K_{i,t} = \alpha_0 + \alpha_1 \log U_{i,t-q} + \alpha_2 \log S_{i,t-q}^{U} + \alpha_3 \log R_{i,t-q} \\ + \alpha_4 \log S_{i,t-q}^{R} + \alpha_5 \log Z_{i,t} + \varepsilon_i \end{aligned} \tag{9}$$

---

[6]   See, for example, Frost and Spence (1995).

where $K_{i,t}$, $U_{i,t-q}$, $S_{i,t-q}^{U}$, $R_{i,t-q}$, $S_{i,t-q}^{R}$, and $Z_{i,t}$ are defined as above; α_1, ..., α_5 are the parameters of interest; α_0 is a constant term and ε_i a stochastic error term. Model (9) has some attractive features. Aside from being easy to estimate, the α are estimates of the elasticities of the increment of economically valuable technological knowledge, $K_{i,t}$, with respect to changes in the respective variables, and these elasticities are constant. But this tractability comes at some cost. The knowledge production function imposes a constant, unitary elasticity of substitution between all input pairs in addition to the constant output elasticities noted above.

We interpret an influence of $U_{i,t-q}$ on $K_{i,t}$ as evidence of intraregional spillovers of local universities in $(t - q, t)$ and an influence of $S_{i,t-q}^{U}$ as evidence of interregional spillovers of universities located outside the region. A lack of significance of α_1 and α_2 would suggest that all production of new knowledge is generated internally to the corporate sector, either with interregional knowledge spillovers originating from firms outside the region if α_4 is significant or without such spillovers if α_4 is not significant.

3   Data, Variable Definition and the Spatial Scale of the Analysis

This paper follows in a tradition that uses patents to measure the outcome of the inventive process, that is knowledge increments. Patents are preferred to innovation counts because it is conceptually more closely related to invention activities[7]. Data on corporate patents of high-technology firms are from the Austrian Patent Office. The patent data file contains information on the application date that can be considered as being relatively close to the date of invention, the name of the assignee(s), the address of the assignee(s), the name of inventor(s), the location of the inventor(s), one or more International Patent Classification (IPC) codes and some information on the technology field of the patent classification.

There is no simple, consistent practice with respect to the names to which corporate patents are assigned. Some patents go only to the assignee. As a consequence, we used the address of the assignee(s) to trace patent activity back to the region of knowledge generation. This approach may be biased in the case of large companies since patents are filed by the headquarter of a company. An extensive effort was made to identify patent receiving subsidiaries and to

---

[7]   See Griliches (1990) for a discussion of the use of patent statistics as economic indicators. It is noteworthy that patents provide only a partial picture of the contributions of university research. But innovation counts are less useful because they measure more aspects of the economic impact of inventive activities rather than the output of the invention process. Innovation counts (generally in terms of improved products on the market) that have been used in most of the US American studies are too far away from the idea of outputs of the inventive process.

redistribute the patents correctly. In the case of multiple assignees located in different regions, we followed the standard procedure of proportionate assignment[8]. We made use of the MERIT concordance table between IPC classes and the industrial ISIC sectors (Verspagen et al. 1994). This table assigns the technical knowledge in the patent classes to the industrial sector best corresponding to the origin of this knowledge. In some cases where the IPC code corresponds to more than one industrial sector, a fractional count was made. Appendix A gives detailed information on the assignment of the patent classes to the industry sectors as used in the paper.

At the sectoral scale, the patent data were aggregated to the two-digit ISIC code level. This is essentially due to data limitations for the explanatory variables in the model, more specifically for the variable on industry R&D investment. Our interest focused on patents in the high-technology sector as an aggregate. The determination of this sector is not unambiguous. We define the high-technology sector to consist broadly of the following six two-digit industries: Computers and Office Machines (ISIC 30); Electronics and Electrical Engineering (ISIC 31-32); Scientific Instruments (ISIC 33); Machinery & Transportation Vehicles (ISIC 29, 34–35); Oil Refining, Rubber & Plastics (ISIC 23, 25); and Chemistry & Pharmaceuticals (ISIC 24). These industries are not equally technology intensive. Some produce more inventions than others, and the propensity to patent these inventions differs between them (see Fischer et al. 1994 for some evidence).

The industries contain most of the three- and four-digit-ISIC categories that are typically classified as high-technology. But at the two-digit ISIC level, it is virtually impossible to designate industries as pure high-technology. To the extent that the sectoral mix in these industries shows some systematic variation over space in its 'pure' high-technology content, our results on the relationship between the increment of economically valuable knowledge and research investment could be affected. But we are confident that we will be able to detect such systematic variation by means of careful specification tests for spatial effects.

We measure industry R&D investment in the high-technology sector using data on R&D expenditures, even though expenditure data might not be a particularly accurate measure of the *real* resources actually used to do R&D (see Alston et al. 1998). The data stem from a R&D survey carried out by the Austrian Chamber of Commerce in 1991. The questionnaire was sent to 5,670 manufacturing firms in Austria. The response rate was 34 percent. The sample can be seen to cover nearly all firms performing R&D activities in Austria. The ZIP code has been used to trace R&D activities back to the origin of knowledge production. The data are broken down by a very specific Industrial Classification System of the Chamber of Commerce that can be converted to the International Standard Classification System only at the fairly broad two-digit ISIC-level.

A major effort was pursued to estimate university research expenditure data for the variable *U*. There are no consolidated research budgets or expense reports available that present data in sufficient detail. We utilised the 1991 survey of the Austrian Federal Ministry for Science and Research to get access to global

---

[8] Note that our dependent variable is, thus, metric.

university research expenditure data. These data include research related basic and on-going operational costs, but not all relevant funding sources. Thus, the data may understate the resources actually used in support of research. But there is no way to overcome this data problem. We proceeded as follows to link university research expenditures to the high-technology industries. First, the global data were broken down by university department on the basis of some simplifying assumptions and a simple disaggregation procedure (see Fischer et al. 2001). Then – using results from Levin et al.'s (1987) survey[9] and Varga's (1998) study in the spirit of Feldman (1994), Audretsch and Feldman (1994), Feldman and Audretsch (1999) – we assigned academic departments and the associated expenditure figures to the six two-digit high-technology industries to which knowledge spillovers from university research may flow. Appendix B shows the match to the two-digit industries. Note that only a smaller set of academic departments produce knowledge relevant to the high-technology sector.

Skilled workers endowed with a high level of human capital are a mechanism through which knowledge externalities materialise. The concentration of skilled labour in one place facilitates flows of information and knowledge because timeliness and face-to-face communication are important for generating new knowledge. To capture such agglomeration externalities (see also Feldman and Florida 1994), we included a location quotient for high-technology employment as a proxy for Z.

The lack of evidence for local geographical spillovers in most US studies is partly – and probably primarily – due to a too high level of spatial data aggregation. In order to overcome this deficiency of previous studies, we have chosen a rather fine level of spatial detail, the scale level of a political district rather than that of a province (Bundesland)[10.] But the price we have to pay for this choice is that this rather fine spatial scale – Austria is divided into 99 political districts – does not support to estimate Equation (9) any more. This is a consequence of the very uneven spatial distribution of universities over the regional system of political districts. There are not enough degrees of freedom or independent variations in the university research expenditure data to allow us to distinguish between inter- and intraregional knowledge spillovers.

One way out of this problem – and the way taken here – is to combine the knowledge spillover aggregates that reflect the pools of intraregional and interregional knowledge spillovers. Let us define, thus, $\Phi_{i,t-q} \equiv (U_{i,t-q} + S^U_{i,t-q})$ and $\Omega_{i,t-q} \equiv (R_{i,t-q} + S^R_{i,t-q})$. Then we get:

---

[9] In Levin et al.'s (1987) survey, R&D managers were asked to indicate on a 7-point Likert scale the relevance of eleven basic and applied fields of science and the importance of external sources of knowledge to technological change in a broad range of manufacturing industries.

[10] This spatial scale is the lowest at which relevant data are available. Political districts – though political-administrative spatial units – are relatively homogeneous in so far that they generally include one larger urban centre and its surroundings.

$$\log K_{i,t} = \beta_0 + \beta_1 \log \Phi_{i,t-q} + \beta_2 \log \Omega_{1,t-q} + \beta_3 \log Z_{i,t} + \xi_i \quad \text{for } i = 1, \ldots, N \quad (10)$$

where β_1, β_2, and β_3 are the parameters of interest; β_0 is a constant and ξ_i a stochastic error term. Φ captures the pool of intra- and interregional university spillovers as an aggregate, and Ω the pool of intra- and interregional knowledge spillovers within the high-technology sector. Specification of the length of the lag relationship has been – and this study makes no exception – largely ad hoc, since past attempts to estimate rather than impose the parameter q have been inconclusive. We follow Verspagen and de Lo (1999) to assume $q = 2$, that is, an average lag of two years for inventions to accompany research expenditures. In our study t refers to the year 1993 and, thus, $t - q$ to 1991.

Finally, it is worth noting that the Cobb-Douglas specification (10) of the regional knowledge production function creates a particular sample selection problem in so far as only observations for which all the variables (dependent and independent) are non-zero can be utilised. Hence, our final data set only includes those political districts for which patents and research expenditures are available. The estimation is carried out with 72 out of 99 observational units for which data are complete. These sample districts represent 100 percent of the university research expenditures (1991); 93.3 percent of the industry R&D activities (1991) and 99.96 percent of the patent applications (1993) in the high-tech sector. The data and specifications used are listed in Appendix C.

4   Estimation Issues

When models such as the Cobb-Douglas versions of (1), (2) and (8) or Equation (10) are estimated for cross-sectoral data on neighbouring spatial units, the lack of independence across these spatial units may lead to spatial dependence (spatial autocorrelation) in the regression equations and, thus, cause serious problems in specifying and estimating the models. In the existing literature, these effects are typically ignored with a few exceptions such as Anselin et al. (1997, 2000). We assess these effects by means of a Lagrange Multiplier [LM] test using six different spatial weights (N, N)-matrices W with $N = 72$ that reflect different a priori notions on the spatial structure of dependence:

- the simple contiguity weights matrix [CONT],
- the inverse distance weights matrix [IDIS1],
- the square inverse distance weights matrix [IDIS2], and
- distance based matrices for 50 km [D50], 75 km [D75] and 100 km [D100] between the administrative centres of the political districts.

This test is used here to assess the extent to which remaining unspecified spatial knowledge spillovers may be present in the knowledge production function model.

Spatial dependence can be incorporated in two distinct ways into the model: as an additional regressor in the form of a spatially lagged dependent variable or in the error structure. The former is referred to as *Spatial Lag Model* and the latter *Spatial Error Model*.

For convenience let be $K = (\log K_{1,t}, \ldots, \log K_{N,t})'$ and $\xi = (\xi_1, \ldots, \xi_N)'$ with $N = 7$. Then the *Spatial Lag Version* of (10) may be expressed in matrix notation as

$$K = \rho W K + X \beta + \xi \tag{11}$$

where K is the (72,1)-vector of observations on the patent variable, $W K$ is the corresponding lag for the (72,72)-weights matrix W, X is a (72, 4)-matrix of observations on the explanatory variables Φ, Ω, Z and a constant term, with matching regression coefficients in the vector β. ξ is a (72, 1)-vector of normally distributed random error terms, with zero mean and constant homoskedastic variance σ^2. ρ is the spatial autoregressive parameter. $W K$ is correlated with the disturbances, even when the latter are i.i.d. Consequently, the spatial lag term has to be treated as an endogenous variable and proper estimation procedures have to account for this endogeneity. Ordinary least squares will be biased and inconsistent due to the simultaneity bias.

The *second way to incorporate spatial autocorrelation* into the regression model (10) is to specify a spatial process for the disturbance terms. The resulting error covariance will be non-spherical, thus, while unbiased, ordinary least squares (OLS) will be inefficient. Different spatial processes lead to different error covariances with varying implications about the range and extent of spatial interaction in the model (Anselin and Bera 1998). The most common specification is a spatial autoregressive process in the error terms that results in the following matrix form of the *spatial error model for regional knowledge production*:

$$K = X \beta + \xi \tag{12}$$

with

$$\xi = \lambda W \xi + \eta \tag{13}$$

that is a linear regression with error vector ξ, where λ is the spatial autoregressive coefficient for the error lag $W \xi$. X is a (72, 4)-matrix of observations on the explanatory variables including a constant term as above, and β a (4, 1)-vector of regression coefficients. The errors ξ are assumed to follow a spatial autoregressive process with autoregressive coefficients, and a white noise error η.

The similarity between the Spatial Error Model (12) – (13) and the Spatial Lag Model (11) for knowledge production complicates specification testing in practice, since tests designed for a spatial lag specification will also have power against a spatial error specification, and vice versa. But as evidenced in a large number of Monte Carlo simulation experiments in Anselin and Rey (1991), the joint use of the Lagrange multiplier tests for spatial lag and spatial error dependence suggested by Anselin (1988) provides the best guidance for model specification. When both tests have high values indicating significant spatial dependence in the data, the one with the highest value (lowest probability) will indicate the correct specification.

5   Empirical Results

Table 1 presents the results of the estimation of the cross-sectional regression of the regional knowledge production function for 72 political districts in Austria and the distance friction parameter[11] $\gamma = 2$. All variables are in logarithms.

We estimated the *Spatial Error Model* version of Equation (10) (see Equations (12)-(13)), and for matters of illustration two special cases of (10). Both assume i.i.d. zero mean error terms. The first, termed *Basic Model,* additionally assumes $\beta_3 = 0$, while the second, termed *Extended Model*, does not, but assumes that knowledge externalities of the *Marshall-Arrow-Romer* and *Isard-Jacobs* type play a decisive role. The results of the *Basic Model* are reported in column 1, the results of the *Extended Model* in column 2 and those of the *Spatial Error Model* in column 3. All estimation and specification tests were carried out with SpaceStat Software (see Anselin 1995).

An influence of Ω on patent activities indicates knowledge production internally to the high-technology industries including geographically mediated spillovers between R&D laboratories. We interpret an influence of Φ on patent activities as evidence of the existence of geographically mediated academic spillovers. The results provide strong further evidence of the empirical relevance of geographic localisation of knowledge spillovers as was indicated, for example, in Jaffe (1989), Ács et al. (1992), Jaffe et al. (1993), Audretsch and Feldman (1994), and Anselin et al. (1997, 2000) for the American case.

All regression models yield highly significant and positive coefficients for both university research and industry R&D spillovers [at $p < 0.01$]. The university research elasticities range in magnitude from 0.128 for the *Basic Model* to 0.130 for the *Spatial Error Model*. The university research effect is much smaller than the industry R&D effect. Knowledge externalities of the *Marshall-Arrow-Romer* and *Isard-Jacobs* type are twice as important as industry R&D effects. For all models, diagnostic tests were carried out for heteroskedasticity, using the White (1980) test. In addition, specification tests for spatial dependence and spatial error were performed, utilising the Lagrange multiplier test. The tests for spatial autocorrelation were computed for the six different spatial weights matrices (CONT, IDIS1, IDIS2, D50, D75 and D100). Only the results for the most significant diagnostics are reported in Table 1.

---

[11] The distance friction parameter has been optimised for the *Basic Model*. The result achieved ($\gamma = 2$) is in accordance with Sivitanidou and Sivitanides (1995). Note that the modelling results obtained are relatively insensitive to the choice of $\gamma \in [1, ..., 4]$.

Table 1   Regression results for (log) patent applications at the level of Austrian political districts ($N = 72$, 1993)

Model	Basic model [OLS]	Extended model [OLS]	Spatial error model [ML]
Constant	0.608*** (0.182)	3.741*** (0.783)	3.315*** (0.764)
Log Φ	0.128*** (0.040)	0.100*** (0.037)	0.130*** (0.037)
Log Ω	0.402*** (0.054)	0.211*** (0.065)	0.213*** (0.064)
Log Z		0.512*** (0.125)	0.438*** (0.121)
Spatial autoregressive coefficient λ			0.366* (0.190)
Adjusted R^2	0.598	0.672	0.699
Multicollinearity condition number	3.978	21.341	21.341
White test for heteroskedasticity	3.210	8.839	
Breusch-Pagan test for heteroskedasticity			2.277
Likelihood ratio test for spatial error dependence			2.863 (D100)
Lagrange multiplier test for spatial error dependence	10.092 (D100)	3.444 (D100)	
Lagrange multiplier test for spatial lag dependence	0.551 (D50)	0.889 (D75)	0.382 (IDIS2)

Notes: Estimated standard errors in parentheses; critical values for the White statistic respectively five and nine degrees of freedom are 11.07 and 16.92 ($p = 0.05$); critical value for the Breusch-Pagan statistic with three degrees of freedom is 7.82 ($p = 0.05$); critical values for Lagrange multiplier lag and Lagrange multiplier error statistics are 3.84 ($p = 0.05$) and 2.71 ($p = 0.10$); critical value for Likelihood ratio-error statistic with one degree of freedom is 3.84 ($p = 0.05$); spatial weights matrices are row-standardised: D100 is a distance-based contiguity for 100 kilometers; D75 a distance-based contiguity for 75 kilometers; D50 a distance-based contiguity for 50 kilometers; IDIS2 inverse distance squared; only the highest values for a spatial diagnostics are reported; * denotes significance at the ten percent level; ** significance at the five percent level and *** significance at the one percent level

The *Basic Model* (column 1) confirms the strong significance of university research and industry R&D spillovers. There is a clear dominance of the coefficient of industry R&D over university research, indicating an elasticity that is about three times higher. There is no evidence of heteroskedasticity, but the Lagrange multiplier test for spatial error dependence strongly indicates misspecification of the model.

When the variable Z is added (see columns 2 and 3), the explanatory power of the regressions is substantially and significantly increased. The model fit increases from 0.60 to 0.70 (measured in terms of adjusted R^2), with a positive and significant effect for the knowledge externalities of the *Marshall-Arrow-Romer* and *Isard-Jacobs* type. Geographically mediated industry R&D and university research spillovers remain positive and significant. But the addition of the variable causes the elasticity of both to drop more or less substantially: industry R&D elasticity from 0.402 to 0.211 and university research elasticity from 0.128 to 0.100. There is no evidence of heteroskedasticity, but the Lagrange multiplier test for spatial error dependence strongly indicates misspecification[12].

The correct interpretation has to be based on the spatial error model that removes any misspecification in the form of spatial autocorrelation. The other results are only reported for completeness' sake. The significant parameter of the error term (λ), the significant value of the Likelihood ratio test in spatial error dependence as well as the missing indication for spatial lag dependence and heteroskedasticity (Breusch-Pagan test, see Breusch and Pagan 1979) are taken as evidence for the correctness of the model. There is little change between the interpretation of the model with and without spatial autocorrelation which is to be expected. The main effect of the spatial error autocorrelation is on the precision of the estimates, but in this case it is not sufficient to alter any indication of significance.

In sum, the maximum likelihood (ML)-estimates in column 3 of Table 1 can be reliably interpreted to indicate the influence of university research on knowledge increment in a political district, not only of university research in the district itself, but also in the surrounding districts. The geographic boundedness of university research spillovers is directly linked to a distance decay effect.

6　Summary and Conclusions

In this paper, we have estimated knowledge spillovers from universities within a knowledge production function framework. The production function approach abandons the details of specific events and concentrates on total output of knowledge generation as a function of industry R&D and university research investment. While this approach is more general than the case study approach, it is also coarser and suffers from a less sound behavioural foundation. Nevertheless, it is currently the only available general way of trying to answer questions about the importance and extent of spatial knowledge spillovers from university research.

The key assumption we made in analysing the link between knowledge spillovers and corporate patent activity was that knowledge externalities are more prevalent in high-technology industries where new – technological and scientific – knowledge plays a crucial role. Knowledge spillovers were captured by means of spatially discounted spillover pools. Our empirical results confirm the presence of geographically mediated knowledge spillovers from university and show that

---

[12] Exogeneity of *R* and *U* were also checked by applying the Durbin-Wu-Hausman test. The null hypothesis of exogeneity was not rejected ($p = 0.22$), suggesting that the single equation estimation methods utilised are correct.

these transcend the geographic scale of the political district. The results also demonstrate that such spillovers follow a clear distance decay pattern, a result that is in accordance with Anselin et al. (1997, 2000) despite differences in research design and context. But these externalities appear to be relatively small in comparison to knowledge externalities of the *Marshall-Arrow-Romer* and *Isard-Jacobs* type. These findings call for policy strategies to facilitate flows of knowledge within Austrian regional systems of innovation.

The findings are also important in that they highlight the relevance of modelling knowledge spillovers in form of spatially discounted external stocks of knowledge. But, some cautionary remarks are in order as well.

- *First*, we have chosen to focus on those districts where patent activity and R&D research in the high-technology industries were observed. This leaves aside the issue of why certain locations have R&D and patent activity and others do not, especially when one of the two is present, but the other not.

- *Second*, we were forced to define the high-technology sector on the basis of two-digit ISIC industries. Many products manufactured by these high-techno-logy industries are medium-tech or even low-tech. This aggregation level evidently masks considerable underlying heterogeneity and may be too crude to capture clearly university research effects. The available industry R&D expenditure data do not match the four- and three-digit ISIC levels. Hence additional progress on the issue will have to await the appearance of better data.

- *Third*, the MAUP problem in spatial analysis teaches us that the results of spatial analytical studies tend to be – more or less – affected by the spatial units of analysis. Thus, the choice of appropriate spatial units is of crucial importance. We have no doubt in mind that political districts qualify as most appropriate units of observation in the Austrian context, not at least because they come rather close to the idea of functional regions. But the choice comes not without some price to be paid: the loss of the ability to clearly distinguish intraregional from interregional spillovers.

- *Fourth*, our knowledge production function framework is comparative-static and hence – as all the previous studies – abstracts from several important dynamic issues. Because changes in knowledge have an impact over many years, there is an intrinsic dynamic relationship between today's research investment and future knowledge generation. There are long, variable and uncertain lags in the interval between the start of a research activity and generating useful knowledge. The problem of the timing of spillovers has – admittedly – not been given adequate attention in our study. Given the diffuse nature of knowledge spillovers and the likely presence of long and variable lags, the assumption of a two year lag may be too crude to adequately capture knowledge spillovers. Much more work needs to be done to estimate rather than to a priori impose the time shape of the lag between the input and the output of the knowledge production process.

- *Fifth*, in the context of our study some major research questions relate to measurement issues. How much of university research in a region is spillable? What is the appropriate size unit (the university institute, the university department or the research group)? These and several other questions are crucial for measuring knowledge spillovers from universities. We have chosen an approach essentially adapted from Varga (1998) to assign academic departments and the associated expenditure figures to the high-technology sector to which knowledge spillovers from university research may flow. The approach is rather heuristic in nature. No doubt that much more research needs to be done to address the above questions in some more depth with the aim to come up with a somewhat more analytical matching procedure.

Overall, one main conclusion of the study is that the spatial dimension of knowledge spillovers is not something that should be disregarded. Even with a less refined model version we were able to describe and illustrate the theoretical and empirical necessity to test for the presence of spatial effects and – when needed – to revise the knowledge production model to include them explicitly. This type of spatial econometric analysis may lead to an increasing understanding of the spatial extent of knowledge spillovers and, thus, provide important empirical support for the theory of endogenous economic growth.

Acknowledgements: An earlier version of this paper was presented at the 17th Pacific Conference of the Regional Science Association International, Portland, Oregon, USA, June 30–July 4, 2001. The research has been supported by grants from the Jubiläumsfonds of the Austrian National Bank (no. 7994) and the Department of Economic Geography and GIScience. The authors are also grateful to Harry Klejian and three anonymous referees for valuable comments.

References

Ács Z., Audretsch D.B. and Feldman M.P. (1994): R&D spillovers and recipient firm size, *The Review of Economics and Statistics* 76 (2), 336-340

Ács Z., Audretsch D.B. and Feldman M.P. (1992): Real effects of academic research: Comment, *American Economic Review* 82 (1), 363-367

Adams J.D. (1993): Science, R&D, and invention potential recharge: U.S. evidence, *American Economic Review* 83 (2), 458-462

Adams J.D. (1990): Fundamental stocks of knowledge and productivity growth, *Journal of Political Economy* 98 (4), 673-702

Alston J.M., Norton G.W. and Pardey P.G. (1998): *Science under Scarcity*, CAB International, New York

Anselin L. (1995): SpaceStat Version 1.90. http://www.spacestat.com

Anselin L. (1988): *Spatial Econometrics*: *Methods and Models*, Kluwer, Dordrecht, Boston

Anselin L. and Bera A. (1998): Spatial dependence in linear regression models with an introduction to spatial econometrics. In: Ullah A. and Giles D. (eds.) *Handbook of Applied Economic Statistics*, Marcel Dekker, New York, pp. 237-289

Anselin L. and Rey S. (1991) Properties of tests for spatial dependence in linear regression models, *Geographical Analysis* 23 (2), 112-131

Anselin L., Varga A. and Ács Z. (2000): Geographic and sectoral characteristics of academic knowledge externalities, *Papers in Regional Science* 79, 435-443

Anselin L., Varga A. and Ács Z. (1997): Local geographic spillovers between university research and high technology innovations, *Journal of Urban Economics* 42, 422-448

Audretsch D.B. and Feldman M.P. (1994): Knowledge spillovers and the geography of innovation and production. Discussion Paper 953, Centre for Economic Policy Research, London

Audretsch D.B. and Feldman M.P. (1996): R&D spillovers and the geography of innovation and production, *American Economic Review* 86 (3), 630-640

Bernstein J.I. and Nadiri M.I. (1988): Interindustry R&D spillovers, rates of return, and production in high-tech industries, *American Economic Review* 78 (2), 429-434

Breusch T. and Pagan A. (1979): A simple test for heteroskedasticity and random coefficient variation, *Econometrica* 47 (5), 1287-1294

Cohen W.M. and Levinthal D.A. (1989): Innovation and learning. The two faces of R&D, *Economic Journal* 99 (397), 569-596

Dosi G. (1988): Sources, procedures and microeconomic effects of innovation, *The Journal of Economic Literature* 26 (3), 1120-1126

Feldman M. (1994): *The Geography of Innovation*, Kluwer, Dordrecht, Boston

Feldman M.P. and Audretsch D.B. (1999): Innovation in cities: Science-based diversity, specialisation and localised competition, *European Economic Review* 43 (2), 409-429

Feldman M.P. and Florida R. (1994): The geographic sources of innovation: Technological infrastructure and product innovation in the United States, *Annals of the Association of American Geographers* 84 (2), 210-229

Fischer M.M., Fröhlich J. and Gassler H. (1994): An exploration into the determinants of patent activities: Some empirical evidence for Austria, *Regional Studies* 28 (1), 1-12

Fischer M.M., Fröhlich J., Gassler H. and Varga A. (2001): The role of space in the creation of knowledge in Austria – An exploratory spatial analysis. In: Fischer M.M. and Fröhlich J. (eds.) *Knowledge, Complexity and Innovation Systems*, Springer, Berlin, Heidelberg, New York, pp. 124-145

Frost M.E. and Spence N.A. (1995): The rediscovery of accessibility and economic potential: The critical issue of self-potential, *Environment and Planning A* 27 (11), 1833-1848

Geroski P. (1995): Markets of technology: Knowledge, innovation and appropriability. In: Stoneman P. (ed.) *Handbook of the Economics of Innovation and Technological Change*, Blackwell, Oxford [UK], Cambridge [MA], pp. 90-131

Griliches Z. (1990): Patent statistics as economic indicators: A survey, *The Journal of Economic Literature* 23, 1661-1707

Griliches Z. (1979): Issues in assessing the contribution of research and development to productivity growth, *Bell Journal of Economics* 10, 92-116

Jaffe A.B. (1989): Real effects of academic research, *American Economic Review* 79, 957-970

Jaffe A.B., Trajtenberg M. and Henderson R. (1993): Geographic localisation of knowledge spillovers as evidenced by patent citations, *Quarterly Journal of Economics* 63 (3), 577-598

Kamien M.I. and Schwartz N.L. (1982): *Market Structure and Innovation*, Cambridge University Press, Cambridge [MA]

Karlsson C. and Manduchi A. (2001): Knowledge spillovers in a spatial context – A Critical review and assessment. In: Fischer M.M. and Fröhlich J. (eds.) *Knowledge, Complexity and Innovation Systems*, Springer, Berlin, Heidelberg, New York, pp. 101-123

Levin R.C., Klevorick A.K., Nelson R.R. and Winter S.G. (1987): Appropriating the returns from industrial research and development, *Brookings Papers on Economic Activity* 1987 (3), 783-820

Mansfield E. (1995): Academic research underlying industrial innovations: Sources, characteristics, and financing, *The Review of Economics and Statistics* 77 (1), 55-65

Mansfield E. (1991): Academic research and industrial innovation, *Research Policy* 20 (1), 1-12

Nelson R.R. (1986): Institutions supporting technical advance in industry, *American Economic Review* 76 (2), 186-189

OECD (1992): *Technology and Economy: The Key Relationships*, Organisation for Economic Co-operation and Development, Paris

Polanyi M. (1967): *The Tacit Dimension*, Doubleday Anchor, New York

Romer P. (1990): Endogenous technological change, *Journal of Political Economy* 98, 72-102

Rosenberg N. (1976): *Perspective on Technology*, Cambridge University Press, Cambridge [MA]

Sivitanidou R. and Sivitanides P. (1995): The intrametropolitan distribution of R&D activities: Theory and empirical evidence, *Journal of Regional Science* 25 (3), 391-415

Varga A. (1998): *University Research and Regional Innovation: A Spatial Econometric Analysis of Academic Technology Transfers*, Kluwer, Dordrecht, Boston

Verspagen B. and De Loo I. (1999): Technology spillovers between sectors and over time, *Technological Forecasting and Social Change* 60, 215-235

Verspagen B., van Moergastel T. and Slabbers M. (1994): MERIT Concordance Table: IPC-ISIC (rev.2), MERIT Research Memorandum 2/94-004, Maastricht Economic Research Institute on Innovation and Technology, University of Limburg

White H. (1980): A heteroskedasticity-consistent covariance matrix estimation and a direct test for heteroskedasticity, *Econometrica* 48 (4), 817-830

Appendix A: Assignment of Patent Classes to the High-Technology Sectors at the Two-Digit ISIC-Level

ISIC category	Industry sector	IPC patent classes
30	Computers & office machinery	B41J, B41L [50%], G06C, G06E, G06F, G06G, G06J, G06K, G06M, G11B, G11C
31–32	Electronics & electrical engineering	A45D [40%], A47J [80%], A47L [40%], A61H [30%], B03C, B23Q [10%], B60Q, B64F [20%], F02P, F21H, F21K, F21L; F21M, F21P, F21Q, F21S, F21V, F27B [10%], G08B, G08G, H01B, H01F, H01G, H01H, H01J, H01K, H01M, H01R, H01S, H01T, H02B, H02G, H02H, H02J, H02K, H02M, H02N, H02P, H03M, H05B, H05C, H05F, H05H, G08C, G09B [50%], H01C, H01L, H01P, H01Q, H03B, H03C, H03D, H03F, H03G, H03H, H03J, H03K, H03L, H04A, H04B, H04G, H04H, H04J, H04K, H04L, H04M, H04N, H04Q, H04R, H04S, H05K
33	Scientific instruments	A61B, A61C, A61D, A61F, A61G [90%], A61H [40%], A61L [60%], A61M, A61N, A62B [50%], B01L, B64F [10%], C12K [25%], C12Q, F16P [60%], F22B [20%], F22D [20%], F22G [20%], F22X [20%], F23N, F23Q [10%], F24F [20%], F41G, G01B, G01D, G01F [60%], G01H, G01J, G01K, G01L, G01M, G01N, G01P, G01R, G01S, G01T, G01V, G01W, G02B, G02C, G02F, G03B, G03C, G03D, G03G, G03H, G04B, G04C, G04F, G04G, G05B, G05C, G05D, G05F, G05G, G06D, G07B, G07C, G07D, G07F, G07G, G09G, G12B, G21F, G21G, G21H, G21K, H05G
29, 34–35	Machinery & transportation vehicles	A01B, A01C, A01D, A01F, A01G [10%], A01J [80%], A01K [30%], A21B, A21C, A21D [30%], A22B [50%], A22C [70%], A23C[10%], A23G [10%], A23N, A23P, A24C, A24D [50%], A43D, A61H [30%], A62B [30%], B01B, B01D, B01F, B01J, B02B [50%], B02C, B03B, B03D, B04B, B04C, B05B [50%], B05C [95%], B05D, B05X [50%], B06B, B07B, B07C, B08B, B09B [25%], B22C [10%], B23Q [70%], B25J, B27J, B28B [60%], B28C [60%], B28D [70%], B29B [80%], B29C [80%], B29D [50%], B29F [80%], B29G [50%], B29H [50%], B29J [40%], B30B, B31B, B31C [90%], B31D [80%], B31F [80%], B41B, B41D, B41F, B41G, B42C [50%], B60C [20%], B65 B, B65C, B65G [40%], B65H, B66B, B66C, B66D, B66F, B66G, B67B [50%],B67C, B67D, C02F [30%], C10F, C12H, C12L, C12M, C13C, C13G, C13H, C14B [50%], C14C [50%],D01B [50%], D01C [50%], D01D [50%], D01F [50%], D01G [50%], D01H [50%], D02D, D02G [50%], D02H [50%], D02J [50%], D03D [50%], D03J, D04B [50%], D04C [50%], D04D [50%], D04G [50%], D04H [50%], D06C, D06F [70%],

ISIC category	Industry sector	IPC patent classes
		D06G, D06H [70%], D21F, D21G, E01B [50%], E01C [50%],E01H [80%], E02D [30%], E03B [30%], E04D [25%], E21B [45%], E21C, E21D [50%], F01B, F01C, F01D, F01K, F01L, F01M, F01N, F01P, F02B, F02C, F02D, F02F, F02G, F02K, F03B, F03C, F03D, F03G, F03H, F04B, F04C, F04D, F04F, F15B, F15C, F15D, F16C, F16J [80%], F16K, F16N, F16T, F23B, F23C, F23D, F23G, F23H, H23J, F23K, F23L, F23M, F23Q [60%], F23R, F24F [80%], F24J [30%], F25B, F25C, F25D, F25J, F26B, F27B [90%], F27D, F28B, F28C, F28D, F28G, F41A, F41B, F41C, F41D, F41F, F41H [50%], F42B, F42C, F42D [50%], G01F [40%], G01G, G21J
23, 25	Oil refining, rubber & plastics	A47G [50%], A47K [40%], A61J [40%], A62B [20%], B29H [50%], B60C [80%], C10B, C10C, C10G, C10L, C10M, D06N [50%], F42D [50%]
24	Chemistry & pharmaceuticals	A01M [20%], A01N, A61J [30%], A61K [95%], A61L [40%], A62D, B09B [75%], B27K [70%], B29B [20%], B29C [20%], B29D [50%], B29F [20%], B29G [50%], B29K, B29L, B41M [15%], B44D [50%], C01B, C01C, C01D, C01F, C01G, C02F [50%], C05B, C05C, C05D, C05F, C05G, C06B, C06C, C06D, C06F, C07B [95%], C07C [95%], C07D [95%], C07F [95%], C07G [95%], C07H [90%], C07J, C07K, C08B, C08C, C08F, C08G, C08H, C08J, C08K, C08L, C09B, C09C, C09D, C09F, C09G, C09H, C09J, C09K, C10H, C10J, C10K, C10N, C11B [50%], C11C [50%], C11D, C12D [90%], C12K [75%], C12N [80%], C12P [50%], C12R [10%], C12S, C14C [50%], E04D [25%], F41H [50%]

Notes: The assignment is based on the MERIT concordance table (Verspagen et al. 1994) between the International Patent Classification (IPC) and the International Standard Industrial Classification of all economic activities (ISIC-rev.2) of the United Nations. The percentages in brackets in the last column of the table give the share of the patents in the IPC-class assigned to the accessory ISIC-category if not all patents in the IPC-class are assigned to the corresponding ISIC-category. A percentage of 100%, for example, therefore means that all patents in the IPC-class are assigned to the corresponding ISIC-category.

Appendix B: Linking Scientific Fields/University Departments to the Two-Digit High-Technology Sectors

ISIC category	Industry sector	Associated scientific fields/university departments
30	Computers & office machinery	Fields connected with information technologies: micro-electronics, automation and robotics, computer sciences, etc.
31–32	Electronics & electrical engineering	Electrical engineering, micro-electronics, technical mathematics, automation and robotics, computer sciences, etc.
33	Scientific instruments	Engineering fields such as mechanical engineering, electrical engineering, micro-electronics, automation and robotics, technical mathematics, computer sciences, physics related fields, medicine related fields, biology related fields, materials sciences, etc.
29,34–35	Machinery & transportation vehicles	Engineering fields including mechanical engineering and electrical engineering, heat science, thermodynamics, material sciences, computer sciences, technical mathematics, astronomy, transport science
23, 25	Oil refining, rubber & plastics	Chemistry related fields including materials sciences, chemical engineering and care chemistry except for certain sectors such as quantum chemistry, biochemistry and geochemistry
24	Chemistry & pharmaceuticals	Chemistry-, pharmaceuticals- and medicine related fields including microbiology, pharmaceutical chemistry, biochemistry, etc.

Source: On the basis of Levin et al. (1987), Feldman (1994); Audretsch and Feldman (1994) and Varga (1998) in the spirit of Feldman and Audretsch (1999); only the most important scientific fields/university departments are listed.

Appendix C: Patent Applications (1993), Industry R&D (1991) and University Research (1991) for 72 Austrian Political Districts

Political district	Patent applications [variable K]	Industry R&D [variable R]	University research and out-of-district access to university research [variable Φ]
Eisenstadt-Umgebung	3.00	35.45	1.24
Neusiedl am See	3.00	7.29	1.38
Oberpullendorf	1.00	3.80	0.52
Klagenfurt (Stadt)	19.50	3.29	36.14
Villach(Stadt)	8.00	16.16	0.13
Hermagor	1.00	0.34	0.09
Sankt Veit an der Glan	1.00	3.16	0.26
Spittal an der Drau	4.00	0.41	0.10
Villach Land	6.50	35.01	0.14
Wolfsberg	2.00	6.24	0.35
Feldkirchen	2.00	0.35	0.20
Krems (Stadt)	2.50	17.74	0.71
Sankt Pölten (Stadt)	7.50	21.34	1.01
Waidhofen (Stadt)	3.00	6.60	0.31
Wiener Neustadt (Stadt)	5.00	14.24	1.65
Amstetten	16.00	87.49	0.37
Baden	27.50	360.98	4.80
Gänserndorf	3.00	14.33	3.19
Korneuburg	12.50	46.70	9.82
Mödling	22.40	213.57	12.97
Neunkirchen	10.00	61.54	1.01
Sankt Pölten (Land)	3.50	4.61	1.45
Scheibbs	1.00	4.98	0.42
Tulln	2.80	34.12	3.29
Waidhofen an der Thaya	1.00	1.20	0.28
Wiener Neustadt (Land)	6.60	11.75	1.55
Vienna-Umgebung	14.60	323.08	25.35
Linz (Stadt)	62.30	1,144.26	218.16
Steyr (Stadt)	28.60	1,123.43	0.36
Wels (Stadt)	12.50	30.87	0.44
Braunau am Inn	8.50	14.73	0.13
Gmunden	19.10	103.77	0.20
Grieskirchen	10.00	49.42	0.24
Kirchdorf an der Krems	12.30	7.21	0.25
Linz-Land	10.70	111.67	2.74
Perg	13.00	26.41	0.44
Ried im Innkreis	5.30	11.96	0.17
Rohrbach	3.00	3.11	0.22
Schärding	5.00	10.34	0.14
Steyr-Land	8.00	10.43	0.28
Vöcklabruck	43.80	318.82	0.20
Wels-Land	5.00	77.04	0.28

Appendix C (*ctd.*)

Political district	Patent applications [variable K]	Industry R&D [variable R]	University research and out-of-district access to university research [variable Φ]
Salzburg (Stadt)	34.30	36.70	117.10
Hallein	8.10	107.28	0.53
Salzburg-Umgebung	23.80	20.92	0.70
Zell am See	5.00	4.57	0.12
Graz (Stadt)	84.30	399.49	1,195.15
Bruck an der Mur	4.30	9.17	1.09
Deutschlandsberg	5.50	93.80	0.97
Feldbach	1.00	2.08	0.81
Fürstenfeld	2.00	12.38	0.61
Graz-Umgebung	8.50	347.15	8.75
Hartberg	1.00	5.53	0.65
Judenburg	12.00	42.26	0.38
Knittelfeld	3.00	20.34	0.48
Leibnitz	4.00	2.23	1.09
Leoben	3.00	5.93	98.51
Liezen	4.00	25.22	0.22
Mürzzuschlag	1.00	9.84	0.55
Voitsberg	10.00	7.88	1.57
Weiz	4.00	123.45	1.68
Innsbruck-Stadt	9.00	5.54	852.03
Innsbruck-Land	29.40	39.07	8.38
Kitzbühel	7.00	15.91	0.18
Kufstein	9.00	329.98	0.25
Lienz	3.00	8.73	0.08
Schwaz	15.00	80.21	2.58
Bludenz	1.00	17.86	0.06
Bregenz	12.00	66.74	0.04
Dornbirn	11.00	146.49	0.04
Feldkirch	14.00	90.23	0.05
Vienna	383.70	6,999.29	3,345.06

Notes: Industry R&D and university research were measured in terms of expenditures, all figures are in millions of 1991 ATS; patent and industry R&D data refer to high-technology industries; university research data include those academic institutes that are expected to be important for the high-technology industries; universities are located in seven political districts: Vienna hosting six universities, Graz (Stadt), Innsbruck (Stadt), Salzburg (Stadt), Linz (Stadt), Klagenfurt (Stadt) and Leoben; all the other political districts have only out-of-district access to university research.

Sources: Patent data were compiled from the Austrian Patent Office database; industry R&D data were compiled from the 1991 industry R&D Survey of the Austrian Chamber of Commerce; university research date were estimated on the basis of information provided by the Austrian Federal Ministry for Science and Research.

11 Patents, Patent Citations and the Geography of Knowledge Spillovers in Europe

with *T. Scherngell and E. Jansenberger*

The main focus in this chapter is on knowledge spillovers between high-technology firms in Europe, as captured by patent citations. High-technology is defined to include the ISIC sectors aerospace (ISIC 3845), electronics-telecommunication (ISIC 3832), computers and office equipment (ISIC 3825), and pharmaceuticals (ISIC 3522). The European coverage is given by patent applications at the European Patent Office that are assigned to high-technology firms located in the EU-25 member states (except Cyprus and Malta), the two accession countries Bulgaria and Romania, and Norway and Switzerland. By following the paper trail left by citations between these high-technology patents, the contribution provides strong evidence for the localisation of knowledge spillovers at two different levels (country, region) in Europe even after controlling for the tendency of inventive activities in the high-technology sector to be geographically clustered. The findings not only indicate that knowledge localisation exists in the aggregate, but also that there are variations of localisation by region.

1 Introduction

As interest in questions of the knowledge economy has grown, knowledge spillovers have received increasing attention in recent years. For the purpose of this paper we use externalities and knowledge spillovers interchangeably to denote the non-pecuniary benefit of knowledge to firms, not responsible for the original investment in the creation of this knowledge. Such spillovers arise when some of the R&D activities have the characteristics of a non-rivalrous[1] good and can not be appropriated entirely.

A fundamental question addressed by empirical research on knowledge spillovers is whether these spillovers are geographically bounded or not (see Karlsson and Manduchi 2001). Most of the studies on this issue thus far have concentrated on the spatial extent of local geographic effects that university research may have on the innovative capacity in a region, both directly and indirectly through its interaction with private sector R&D efforts. The studies vary somewhat in terms of research design, but they all find a strong and positive

---

[1] Non-rivalry implies that the knowledge in question can be used many times and in many different processes without losing its value.

relationship between innovative activity and both industry R&D and university research at the state level in the USA. But the situation is different in terms of the significance of a *local* geographic spillover effect. Overall, the evidence is non-existent, weak or mixed, and only pertaining to a few individual sectors (Anselin et al. 1997). This may be due to the fact that knowledge spillovers are measured indirectly rather than directly.

The only *direct* evidence we have for localised knowledge spillovers is based on Jaffe's et al. (1993) pioneering analysis on patent citations. Our study lies in this research tradition[2] and takes patent citations as a proxy for knowledge spillovers to test for spillover localisation in Europe. We are particularly interested in knowledge spillovers between high-technology firms. High-technology is defined in our context to include the ISIC-sectors (ISIC Rev. 2) pharmaceuticals (ISIC 3522), computers and office equipment (ISIC 3825), electronics-telecommunication (ISIC 3832), and aerospace (ISIC 3845). Though some firms may choose not to patent inventions, patenting in high-technology industries is commonly practiced and indeed a vital component of maintaining technological competitiveness. The European coverage of our study is given by patent applications at the European Patent Office (EPO) that are assigned to high-technology firms located in the EU-25 member states[3], the two accession countries Bulgaria and Romania, and Norway and Switzerland. Space is considered in a discrete representation of 188 regions[4].

The rest of the paper is structured as follows. The next section explains the nature of patents in some more detail and illustrates those inventive activities in the high-technology sector in Europe to be geographically clustered. Knowledge flows are notoriously difficult to measure. Following Jaffe, Trajtenberg and co-authors (see Jaffe and Trajtenberg 2002) we use patent citation data as an indicator for a specific type of knowledge spillovers between inventors. Section 3 considers more carefully how citations might be used to infer spillovers. The section, moreover, illustrates how patent citations between high-technology firms are spread across Europe and tend to be geographically clustered. Section 4 follows the pioneering methodology developed in Jaffe et al. (1993) and compares the extent to which actual citations are disproportionately located in space, relative to a distribution of control patents that have the same temporal and technological characteristics. This comparison allows us to control for any technology-based clustering of inventive activity, which may otherwise confound any inference drawn from co-location of citations. Section 5 concludes with a summary of our main findings and some suggestions for future work.

---

[2] See Almeida (1996); Almeida and Kogut (1999); Hicks et al. (2001); Maurseth and Verspagen (2002); Agrawal et al. (2003); Singh (2003), and Verspagen and Schoenmakers (2004) for examples of this research tradition.

[3] Except Cyprus and Malta.

[4] For the definition of the regions see Appendix A.

2   Patents and Patent Data

Patents have long been recognised as an important and fruitful source of data for the study of innovation and technological change (see Griliches 1990 for a survey of the use of patent statistics). A patent is a property right awarded to inventions for the commercial use of a newly invented device. An invention to be patented has to satisfy three patentability criteria. It has to be *novel* and *non-trivial* in the sense that it would not appear obvious to a skilled practitioner of the relevant technology, and it has to be *useful*, in the sense that it has potential commercial value. If a patent is granted, an extensive public document is created. The document contains detailed information about the technology of the invention, the inventor, the assignee that owns the patent rights, and the technological antecedents of the invention. Because patent documents record the residence of the inventors they are an important resource for analysing the spatial extent of knowledge spillovers, as captured by patent citations.

Patent related data have, however, two important limitations. First, the range of patentable inventions constitutes only a subset of all R&D outcomes, and second, patenting is a strategic decision and, thus, not all patentable inventions are actually patented. As to the first limitation, purely scientific advances devoid of immediate applicability as well as incremental technological improvements which are too trite to pass for discrete, codifiable inventions are not patentable. The second limitation is rooted in the fact that it may be optimal for inventors *not* to apply for patents even though their inventions would satisfy the criteria for patentability (Trajtenberg 2001). Inventors balance the time and expense of the patent process, and the possible loss of secrecy which results from patent publication, against the protection that a patent potentially provides to the inventor (Jaffe 2000). Therefore, patentability requirements and incentives to refrain from patenting limit the scope of our analysis based on patent data.

Patents from different national patent offices are not comparable to each other because of different patent breadth, patenting costs, approval requirements, citation practices and enforcement rules across Europe. This makes patent data from the European Patent Office (EPO) rather than national patent offices a natural choice for our study[5]. Our data source is the EPO database. The data on patent applications that we use in this study were drawn from the universe of European patents. By *European* patents, we mean patents assigned to corporations located in Europe, regardless of the nationality of the inventors. Our sample of patents is limited to those that are related to inventions in the high-technology industries or in other words to those patents assigned to patent classes which match the high-technology sector, at the four-digit level of the International Standard Industrial Classification, ISIC Rev. 2. We used MERIT's concordance

---

[5]   At present national systems and the European system function in parallel, though inventors tend to be making increasing use of the European system. This is especially true for inventors in smaller European countries looking for wider geographical protection for their inventions. But nevertheless it should be noted that patent data from the EPO cover only a subsample of patents applied for in Europe.

table (see Verspagen et al. 1994) between the four-digit ISIC-sectors and the 628 patent subclasses[6] of the International Patent Code (IPC) classification to identify the high-technology patents from the universe of European patent applications. The patent subclasses associated with the four digits ISIC sectors are outlined in Appendix B.

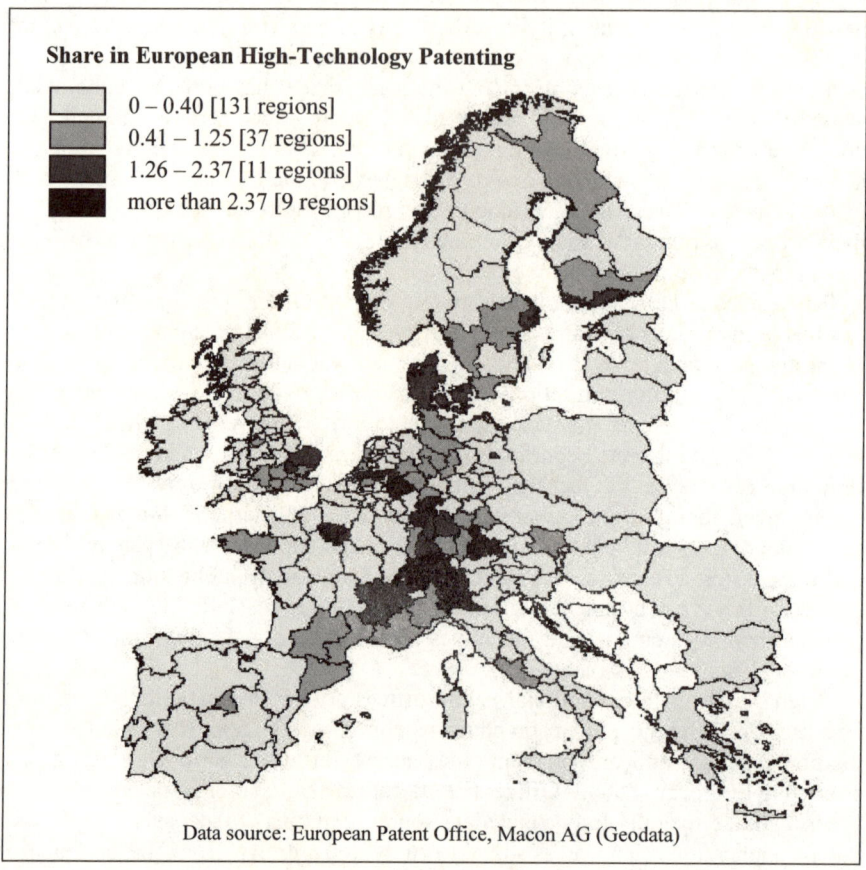

Share in European High-Technology Patenting

- 0 – 0.40 [131 regions]
- 0.41 – 1.25 [37 regions]
- 1.26 – 2.37 [11 regions]
- more than 2.37 [9 regions]

Data source: European Patent Office, Macon AG (Geodata)

Figure 1   Geographic distribution of high-technology EPO-patents across European regions (1985-2002), measured in terms of shares in European high-technology patenting

Our database contains all the high-tech patents applied at the EPO between 1985 and 2002, totalling 177,424 patents. Each patent application produces a highly structured public document containing detailed information on the invention itself,

---

[6]   The IPC system is an internationally agreed, clear-cut non-overlapping hierarchical classification system that consists of five hierarchical levels. At the third level 628 subclasses are distinguished.

the technological area to which it belongs, the inventor and her/his address, and the organisation to which the inventor assigns the patent property right. By nature of the research question, we are interested in the geographical location of the inventor rather than the applicant and, thus, use the postal code of the inventor address for tracing inventive activities back to the region of knowledge production.

For representing geographic space we use 188 regions that cover the EU-25 countries (except Cyprus and Malta), Bulgaria, Romania, Norway and Switzerland. Their definition is based on the Nomenclature des Unites Territoriales Statistiques (NUTS). The regions are essentially in line with the NUTS-2 level of the regional classification in the case of Austria, Belgium, Germany, Finland, France, Italy, The Netherlands, Portugal, Spain, Sweden and UK, and in line with the NUTS-0 level in all other cases. See Appendix A for the exact definition of the regions.

Table 1   The Top-25 European regions in high-technology patenting (1985-2002)

NUTS code	Region	Share in European patenting
FR10	Île-de-France	9.21
DE21	Oberbayern	6.76
CH00	Switzerland	4.49
NL41	Noord-Brabant	4.46
DE71	Darmstadt	3.52
DEA1	Düsseldorf	2.87
IT20	Lombardia	2.83
DEA2	Köln	2.72
DEB3	Rheinhessen-Pfalz	2.41
SE01	Stockholm	2.28
FR71	Rhône-Alpes	2.18
DK00	Denmark	1.93
DE11	Stuttgart	1.91
FI16	Uusimaa	1.83
DE12	Karlsruhe	1.78
UKH1	East Anglia	1.76
DE30	Berlin	1.65
UKI2	London Region	1.52
DE13	Freiburg	1.45
DE25	Mittelfranken	1.31
UKH2	Bedfordshire & Hertfordshire	1.31
UKJ2	Surrey	1.27
DEF0	Schleswig-Holstein/Hamburg	1.22
FR82	Provence-Côte d'Azur	1.03
UKH3	Essex	1.02
Sum		64.72

Data source: European Patent Office

Figure 1 shows that inventive activities of high-technology industries, as measured in terms of EPO-patent activities (1985-2002), are unevenly distributed across Europe. High patenting activity is located in the Île-de-France (9.21 percent of European patenting), followed by Oberbayern (6.76 percent), Switzerland (4.49 percent), Noord-Brabant (4.46 percent) and Darmstadt (3.52 percent). The Top-25 regions (see Table 1) account about two thirds of the total number of high-technology patents, which indicates a high geographic concentration in only a few European regions. It is notable that Eastern European regions and Southern European regions (except Northern Italy) display very little patent activity.

3   Knowledge Spillovers, Patent Citations and Data

Patent documents include references or citations to patents. These citations open up the possibility of tracing multiple linkages between inventions, inventors, firms and locations. In particular, patent citations enable us to analyse the geographical extent of spillovers. There are, however, also some serious limitations to the use of patent citation data. Patent citations capture only those spillovers which occur between patented inventions, and, thus, underestimate the actual extent of knowledge spillovers. Other channels of knowledge transfer – for example, transfer of knowledge embodied in skilled labour, knowledge transfer between customers and suppliers, knowledge exchange at conferences and trade fairs – are not captured by patent citations. Patent citations do not always represent what we typically think of as knowledge spillovers. Some citations may represent only indirect knowledge spillovers since the patent examiner added them. This noise creates a bias against finding spillovers. Fortunately, bias in this direction is a problem of power which can be overcome with a sufficiently large sample size (Thompson 2003).

In constructing the patent citation data set that forms the basis of our study we begin with the full set of issued patents that have their application year between 1985 and 2002. There are 177,424 high-technology patents. We then discard all patents that have not received any citations, since our study is using citations as a proxy for knowledge spillovers. We don't believe that this elimination results in a selection bias since we are interested in comparing the fraction of citations that are from the same location as the original patent, a measure that is conditional on there being citations. Consequently, 42.9 percent of the patents are discarded, leaving 101,247 patents which generate 210,667 citations.

The observation of citations is evidently subject to a truncation bias because we observe citations for only a portion of the life of an invention, with the duration of that portion varying across patent cohorts. This means that patents of different ages are subject to different degrees of truncation. To overcome this problem at least in part we have identified all the pairs of cited and citing patents where citations to a patent are counted for a window of five years following its issuance[7]. The analysis is, thus, confined to 1985-1997 in the case of cited patents while citing patents appearing in 1990-2002 are taken into account. This process reduces the number of patents to 69,814 that generate 155,462 citations. Next, we discard 36.8 percent of those citations for which the citing patent is a self-citation[8], because self-citations do not represent knowledge spillovers in the sense of externalities. This leaves us with 98,191 citations or observations that link a citing patent to a cited patent.

6,000 citations received
3,000 citations received

Data source: European Patent Office; visualisation tool: Borgatti et al. (1999)

Figure 2   Knowledge flows between European regions, as captured by interfirm patent citations in the high-technology sector (1985-2002)

---

[7]   The mean citation lag of all 210,667 citations is 4.6 years, with some sectoral differences: pharmaceuticals (4.4 years), computers and office equipment (4.4 years), electronics-telecommunication (4.7 years) and aerospace (5.4 years).

[8]   We consider assignee matches as self-cites. This is in agreement with most citation-based empirical research.

The original data come in form of citations *made* (that is, each patent lists references to previous patents) while for identifying the knowledge flows one needs a list of cited and citing patent applications. To obtain the citations received by any *one* patent issued in year *t*, one needs to search the references made by all patents applied after year *t*. This requires in fact fast access to all citation data in a way that permits efficient research and extraction of citations not by the patent number of the *citing* patent, but by the patent number of the *cited* patent.

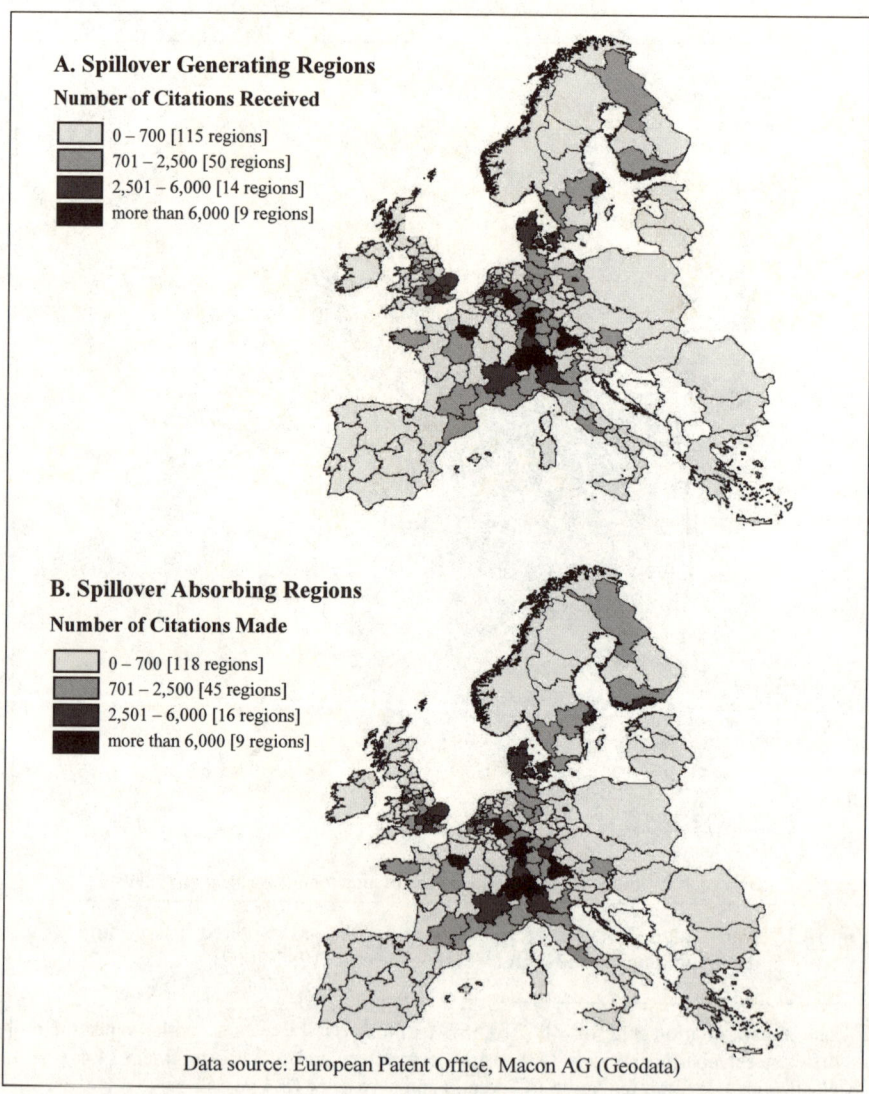

A. Spillover Generating Regions

Number of Citations Received

- 0 – 700 [115 regions]
- 701 – 2,500 [50 regions]
- 2,501 – 6,000 [14 regions]
- more than 6,000 [9 regions]

B. Spillover Absorbing Regions

Number of Citations Made

- 0 – 700 [118 regions]
- 701 – 2,500 [45 regions]
- 2,501 – 6,000 [16 regions]
- more than 6,000 [9 regions]

Data source: European Patent Office, Macon AG (Geodata)

Figure 3   Knowledge spillovers between high-technology firms (1985-2002): A. Spillover generating regions and B. Spillover absorbing regions

The unit of analysis is the dyad 'cited patent-citing patent'. A single originating patent, for example, that has two inventors and is cited by three subsequent patents will generate six unique observations. Each patent is assigned to one of the 188 regions based on the home address of the inventors as reported in the patent document. The 98,191 observations are illustrated in Figure 2. The nodes represent the 188 European regions, their size is relative to their spillover generating power measured in terms of citations received.

Figure 3 complements this picture illustrating the geography of knowledge spillovers across Europe. This figure classifies the 188 regions according to their spillover generating power (measured in terms of citations received; see Figure 3A) and their spillover absorbing power (measured in terms of citations made; see Figure 3B)[9]. Both figures pinpoint to a centre-periphery pattern that is in close line with the pattern of patenting activity across Europe as observed in Figure 1.

4   Testing for Geographic Localisation

Patents linked by citations not only share a technology, but they are also often developed by inventors working in a common industry. Patents linked by a citation are, therefore, much more likely to share a geographic location than a pair of patents drawn at random from the entire pool of patents. To control for the tendency of inventive activities to be geographically clustered – as observed in Section 2 – we follow the case-control matching approach pioneered by Jaffe et al. (1993).

The essence of this case-control approach is to compare citing patents with control patents in terms of the frequency with which each is located in the same region as the originating patent. A finding of a disproportionate number of co-located citations relative to co-located control patents is interpreted as evidence of localised knowledge spillovers. The reason for utilising controls is that patent citations will tend to be co-located with the original inventions even in the absence of knowledge spillovers when inventive activity in particular industries is clustered geographically (Agrawal et al. 2003). Therefore the spillover effect is identified in our study as the extent of co-location which exists over and above what we would expect given the geographic concentration of inventive activity by the high-technology sector.

More formally, let P(citation) be the probability that the originating patent and the citing patent are geographically matched, and P(control) be the corresponding probability for the originating patent-control patent match. Assuming binomial distributions, we test the null hypothesis

H_0: P(citation) = P(control)

versus the alternative hypothesis

---

[9]   The Top-25 regions are listed in Appendix C.

H$_a$: P(citation) > P(control)

using the test statistic

$$t = \frac{P(\text{citation}) - P(\text{control})}{\sqrt{\frac{1}{n}\left\{P(\text{citation})\left[1 - P(\text{citation})\right] + P(\text{control})\left[1 - P(\text{control})\right]\right\}}}$$

where P(citation) and P(control) are the sample proportion estimates of P(citation) and P(control). This statistic tests for the difference between two independently drawn binomial proportions. A positive significant value of Student's t indicates support of the proposition that knowledge flows, proxied by patents cited by the originating patents, are geographically more located than expected.

We use the following procedure to construct the set of control patents. A control patent is selected for each originating patent that matches the citing patent on the following two dimensions: application year and technology classification. Having generated the set of patents with the same application year and the same original three-digit IPC classification code as the citing patent, we identify the patent in the set that has the closest application date to the citing patent. Next, we confirm that the patent does not cite the original patent. If it does, the patent is removed from the set of potential control patents and the next best control patent is selected. Finally, if there are no patents that match the citing patent in at least the application year and the original IPC-classification without citing the original patent, the observation (originating patent) is removed from the data set.

Table 2   Descriptive statistics

Sample	Patents (number)	Total citations	Self-cites[a] (%)	Same patent class[b] (%)	Mean citations received[c]	Average citation lag[d]
1990 Cohort of originating patents	2,118	2,362	31.75	76.54	1.94	4.45
1995 Cohort of originating patents	1,814	2,387	31.84	77.13	1.95	4.57

[a] A self-citation is defined as a citing patent assigned by its inventor to the same party as the originating patent. [b] Comparison is at the three-digit level of the IPC classification. [c] For those patents receiving any citations. [d] Application year of the citing patent minus application year of the originating patent.

We consider two cohorts of originating patents, and corresponding sets of citing patents and control patents to test for spillover localisation. One consists of 1990 patent applications and the other of 1995 applications drawn from our patent database described in Section 2. Table 2 briefly describes these two samples. The 1990 cohort of originating patents contains 2,118 patents that have received a total of 2,362 citations (including self-citations) and 1,410 citations excluding self-citations by the end of 1995. The 1995 cohort of originating patents contains 1,814 patents that have received a total of 2,387 citations (including self-citations) and 1,366 citations excluding self-citations by the end of 2000.

The results of the case-control tests are provided in Table 3 for both cohorts of originating patents. Localisation effects are reported at two spatial levels: the regional and the country level of analysis. *Number of citations* corresponds to the number of citations cited by the originating cohort of patents. *Overall citation matching*, *citation matching excluding self-cites* and *control matching* are the percentage of cited patents (with and without self-citations) and controls that belong to the same geographic location as the originating patent. The *t*-statistic tests the equality of the control proportions and the citation proportions excluding self-citations.

Let us focus first on the 1990 results as displayed in the first column of Table 3. Starting with the country match, we find that citations *including self-citations* are intranational about 38 percent points more often than the controls. Excluding self-citations cuts this difference roughly in half. The remaining difference between the citations excluding self-citations and the controls is strongly significant statistically. Looking at the 1990 results for regions, we find that citations of patents come from the same region about 37 percent of the time. Excluding self-citations, however, makes a big difference. The proportions are cut to 13.7 percent. The matching frequency excluding self-citations is significantly greater than the matching control proportion.

Table 3 Geographic matching fractions[a]

	1990 *Originating cohort*	1995 *Originating cohort*
Number of citations		
incl. self-cites	2,362	2,387
excl. self-cites	1,410	1,366
Matching by country		
Overall citation matching (%)	60.1	61.2
Citation matching excl. self-cites (%)	36.6	35.9
Control matching (%)	21.9	25.4
t-statistic (excl. self-cites)	8.68	6.01
	(*p* = 0.00)	(*p* = 0.00)
Matching by region		
Overall citation matching (%)	36.7	37.0
Citation matching excl. self-cites (%)	13.7	14.8
Control matching (%)	5.2	5.4
t-statistic (excl. self-cites)	7.91	8.27
	(*p* = 0.00)	(*p* = 0.00)

[a] The *t*-statistic tests equality of the citation proportion excluding self-citations and the control proportion. See text for details.

The results for patent citations of the 1995 patents are similar (see the second column in Table 3). For both cohorts of originating patents and for both geographical levels, the patent citations are quantitatively and statistically significantly more localised than the controls. The citation matching percentages slightly rise at the regional level from 13.7 percent in 1990 to 14.8 percent in 1995, but slightly decrease at the country level from 36.6 percent to 35.9 percent. It is impossible, however, to tell from this comparison whether this represents a real change, or whether it is the result of differences in average citation lags[10].

The results on the extent of localisation can be summarised as follows. For citations observed by 1,410 of the 1990 originating cohort of patents, there is a clear pattern of localisation at the regional and country levels. Citations are about seven times more likely to come from the same region than control patents, 2.6 times more likely excluding self-citations. They are 2.7 times more likely to come from the same country as the originating patents, and 1.7 times excluding self-citations. For citations of 1995 originating patents, the same pattern emerges. All these differences are statistically significant at a level much less than one percent.

Table 4   Regional variations in localisation: Tests in selected regions

	Number of citations (excl. self-cites)		Citation matching (%)		Control matching (%)		t-statistic[a]	
	1990	1995	1990	1995	1990	1995	1990	1995
Île-de-France	130	197	27.9	28.4	13.9	8.6	**3.30** (0.000)	**6.05** (0.000)
Oberbayern	82	88	12.1	10.2	2.4	2.4	**2.22** (0.009)	**1.51** (0.037)
Switzerland	73	81	17.8	28.3	9.5	6.1	**1.51** (0.046)	**3.81** (0.000)
Lombardia	68	43	26.4	16.2	7.3	11.6	**3.38** (0.000)	0.70 (0.242)
Noord-Brabant	65	14	24.6	7.1	13.8	7.1	**1.72** (0.044)	0.00 (0.500)
Darmstadt	53	76	11.3	28.9	0.2	3.9	**1.93** (0.029)	**3.95** (0.000)
Köln	38	47	10.5	8.5	2.6	0.0	1.35 (0.091)	**2.06** (0.041)
Bedfordshire	36	13	46.1	23.0	5.5	0.0	**3.21** (0.001)	**1.89** (0.042)
Düsseldorf	28	33	21.4	18.1	3.5	9.0	**2.42** (0.011)	**1.78** (0.022)

[a] results significant at the 5 percent level of significance are in bold.

However, localisation of knowledge spillovers is not a universal phenomenon. European regions reveal different patterns in the local diffusion of knowledge

---

[10] The average citation lag for the 1990 (1995) cohort of originating patents is 4.45 (4.57) compared to 4.14 (4.51) for the corresponding control patents.

externalities. Table 4 presents the results for the Top-8 European regions in high-technology patenting plus Bedfordshire. For the samples, there are significantly higher proportions of citation matches than control matches (except Noord-Brabant in 1995) indicating localisation effects. Results that are significant at the 0.05 level or better are given in bold. These results indicate quite strongly that knowledge is localised at the regional level. In 1995 Île-de-France shows by far the strongest localisation effect. The results for the German regions (Darmstadt, Düsseldorf and Oberbayern), Switzerland and Bedfordshire are also significant in 1990 and 1995[11].

In Table 5 we test whether the degree of knowledge localisation is significantly different across regions. We use the Top-3 regions in high-technology patenting, Île-de-France, Oberbayern and Switzerland, for the comparison. The results show that knowledge spillovers are significantly more localised in the Île-de-France than in any other region though the other two regions also evidence considerable localisation.

Table 5 Test for regional variations in localisation: Results of t-test[a]

	Île-de-France		Oberbayern		Switzerland	
	1990	1995	1990	1995	1990	1995
Île-de-France	–	–	–4.51	–5.55	–3.44	–4.17
Oberbayern	4.51	5.55	–	–	–2.21	–2.53
Switzerland	3.44	4.17	2.21	2.53	–	–

[a] All figures are t-statistics for differences in regional localisation, significant at the 0.05 percent level.

5 Summary and Conclusions

Localisation of knowledge spillovers is implicit in most theories of new economic growth, but rarely studied empirically. In this paper we have analysed patent citation data pertaining to high-technology firms in Europe to test the extent of localisation of knowledge spillovers. As described in the previous section, we compared the probability that citing patents are from the same location as the originating patent with the probability that control patents selected to match the citing patents in terms of timing and technology classification are from the same location as the originating patent.

The results strongly support the hypothesis that spillovers are geographically localised. The proportion of citing patents that match the location of their originating patents is significantly greater than that of control patent location matches at both spatial levels: the country and the region level. The t-statistics, which tests the equality of the proportion of citing-original versus control-original location matches, are large, with $p=0.000$. It is also interesting to note that

---

[11] An examination of the citation data with self-cites reveals that localisation may often be driven by self-citations.

spillover localisation is specific to certain regions and that the degree of localisation is significantly different across regions.

Overall considered, the results support the conclusion that regional and national systems of innovation matter (see Fischer 2001). This is a conclusion that has important policy implications. European regional cohesion appears to be at stake, especially – but not exclusively – because of the localised nature of knowledge flows.

Acknowledgements: The authors gratefully acknowledge the grant no. 11329 provided by the Jubiläumsfonds of the Austrian National Bank. The authors are also grateful to Bernd Bettels (EPO) for supplying the data and to Katarina Kobesova (Institute for Economic Geography and GIScience) for developing the patent citation database.

References

Agrawal A., Cockburn I.M. and McHale J. (2003): Gone but not forgotten: Labor flows, knowledge spillovers and enduring social capital, NBER Working Paper 9950, Cambridge [MA]

Almeida P. (1996). Knowledge sourcing by foreign multinationals: Patent citation analysis in the U.S. semiconductor industry, *Strategic Management Journal* 17, 155-165

Almeida P. and Kogut B. (1999): Localisation of knowledge and the mobility of engineers in regional networks, *Management Science* 45 (7), 905-917

Anselin L., Varga A., and Ács Z. (1997): Local geographic spillovers between university research and high technology innovations, *Journal of Urban Economics* 42 (3), 422-448

Audretsch D.B. and Feldman M.P. (1996): R&D spillovers and the geography of innovation and production, *American Economic Review* 86 (3), 630-640

Borgatti S.P., Everett M.G. and Freeman L.C. (1999): *Ucinet for Windows*: *Software for Social Network Analysis, User's Guide*, Analytical Technologies, Harvard

Fischer M.M. (2001): Innovation, knowledge creation and systems of innovation, *The Annals of Regional Science* 35 (2), 199-216

Fischer M.M. and Varga A. (2003): Production of knowledge and geographically mediated spillovers from universities, *The Annals of Regional Science* 37 (2), 303-323

Griliches Z. (1992): The search for R&D spillovers, *Scandinavian Journal of Economics* 94, Supplement, 29-47

Griliches Z. (1990): Patent statistics as economic indicators: A survey. *The Journal of Economic Literature* 28 (4), 1661-1707

Hatzichronoglou T. (1997): Revision of the high-technology sector and product classification, STI Working Paper 1997/2, Organisation for Economic Co-operation and Development, Paris

Hicks D., Breitzman T., Olivastro D. and Hamilton K. (2001): The changing composition of innovative activity in the USA – A portrait based on patent analysis, *Research Policy* 30 (4), 681-703

Jaffe A.B. (2000): The U.S patent system in transition: Policy innovation and the innovation process, *Research Policy* 29 (5), 531-557

Jaffe A.B. and Trajtenberg M. (eds.) (2002): *Patents, Citations & Innovations. A Window on the Knowledge Economy*, MIT Press, Cambridge [MA] and London [UK]

Jaffe A.B., Trajtenberg M., and Henderson R. (1993): Geographic localisation of knowledge spillovers as evidenced by patent citations, *Quarterly Journal of Economics* 108 (3), 577-598

Karlsson C. and Manduchi A. (2001): Knowledge spillovers in a spatial context – A critical review and assessment. In: Fischer M.M. and Fröhlich J. (eds.) *Knowledge, Complexity and Innovation Systems,* Springer, Berlin, Heidelberg, New York, pp. 101-123

Maurseth P.B. and Verspagen B. (2002): Knowledge spillovers in Europe: A patent citation analysis, *Scandinavian Journal of Economics* 104 (4), 531-545

Paci R. and Batteta E. (2003): Innovation networks and knowledge flows across the European regions, WP CRENoS 13/03, University of Cagliari and CRENoS

Romer P.M. (1990): Endogenous technological change, *Journal of Political Economy* 98 (5), 71-102

Singh J. (2003): Multinational firms and international knowledge diffusion: Evidence using patent citation data, Working Paper, Harvard Business School

Thompson P. (2003): Patent citations and the geography of knowledge spillovers: What do patents examiners know? Manuscript, Florida International University

Trajtenberg M. (2001): Innovation in Israel 1968-1997: A comparative analysis using patent data, *Research Policy* 30 (3), 363-389

Verspagen B. and Schoenmakers W. (2004): The spatial dimension of patenting by multinational firms in Europe, *Journal of Economic Geography* 4 (1), 23-42

Verspagen B., van Moergastel T. and Slabbers M. (1994): MERIT Concordance Table: IPC-ISIC (rev.2), MERIT Research Memorandum 2/94-004, Maastricht Economic Research Institute on Innovation and Technology, University of Limburg

Appendix A: List of Regions Used in the Study

Country	Nuts-code	Region	Country	Nuts-code	Region
Austria	AT11	Burgenland		DEE1	Dessau
	AT12	Niederösterreich/Wien		DEE2	Halle
	AT21	Kärnten		DEE3	Magdeburg
	AT22	Steiermark		DEF0	Schleswig-Holst./Hamburg
	AT31	Oberösterreich		DEG0	Thüringen
	AT32	Salzburg	Denmark	DK00	Denmark
	AT33	Tirol	Estland	EE00	Estland
	AT34	Vorarlberg	Finland	FI13	Itä-Suomi
Belgium	BE10	Région Bruxelles-Capital		FI14	Väli-Suomi
	BE21	Antwerpen		FI15	Pohjois-Suomi
	BE22	Limburg (B)		FI16	Uusimaa
	BE23	Oost-Vlaanderen		FI17	Etelä-Suomi
	BE24	Vlaams Brabant	France	FR10	Île-de-France
	BE25	West-Vlaanderen		FR21	Champagne-Ardenne
	BE31	Brabant Wallon		FR22	Picardie
	BE32	Hainaut		FR23	Haute-Normandie
	BE33	Liège		FR24	Centre
	BE34	Luxembourg (B)		FR25	Basse-Normandie
	BE35	Namur		FR26	Bourgogne
Bulgaria	BG00	Bulgaria		FR30	Nord-Pas-de-Calais
Czech Republic	CZ00	Czech Republic		FR41	Lorraine
Germany	DE11	Stuttgart		FR42	Alsace
	DE12	Karlsruhe		FR43	Franche-Comté
	DE13	Freiburg		FR51	Pays de la Loire
	DE14	Tübingen		FR52	Bretagne
	DE21	Oberbayern		FR53	Poitou-Charentes
	DE22	Niederbayern		FR61	Aquitaine
	DE23	Oberpfalz		FR62	Midi-Pyrénées
	DE24	Oberfranken		FR63	Limousin
	DE25	Mittelfranken		FR71	Rhône-Alpes
	DE26	Unterfranken		FR72	Auvergne
	DE27	Schwaben		FR81	Languedoc-Roussillon
	DE30	Berlin		FR82	Provence-Côte d'Azur
	DE40	Brandenburg	Greece	GR00	Greece
	DE71	Darmstadt	Hungary	HU00	Hungary
	DE72	Gießen	Ireland	IE00	Ireland
	DE73	Kassel	Italy	IT11	Piemonte
	DE80	Mecklenburg-Vorpommern		IT12	Valle d'Aosta
	DE91	Braunschweig		IT13	Liguria
	DE92	Hannover		IT20	Lombardia
	DE93	Lüneburg/Bremen		IT31	Trentino-Alto Adige
	DE94	Weser-Ems		IT32	Veneto
	DEA1	Düsseldorf		IT33	Friuli-Venezia Giulia
	DEA2	Köln		IT40	Emilia-Romagna
	DEA3	Münster		IT51	Toscana
	DEA4	Detmold		IT52	Umbria
	DEA5	Arnsberg		IT53	Marche
	DEB1	Koblenz		IT60	Lazio
	DEB2	Trier		IT71	Abruzzo
	DEB3	Rheinhessen-Pfalz		IT72	Molise
	DEC0	Saarland		IT80	Campania
	DED1	Chemnitz		IT91	Puglia
	DED2	Dresden		IT92	Basilicata
	DED3	Leipzig		IT93	Calabria

Appendix A (*ctd.*)

Country	Nuts-code	Region	Country	Nuts-code	Region
	ITA0	Sicilia	United Kingdom	UKC1	Tees Valley & Durham
	ITB0	Sardegna		UKC2	Northumberland & Wear
Lithuania	LT00	Lithuania		UKD1	Cumbria
Luxembourg	LU00	Luxembourg		UKD2	Cheshire
Latvia	LV00	Latvia		UKD3	Greater Manchester
Netherlands	NL11	Groningen		UKD4	Lancashire
	NL12	Friesland		UKD5	Merseyside
	NL13	Drenthe		UKE1	East Riding & Lincolnsh.
	NL21	Overijssel		UKE2	North Yorkshire
	NL22	Gelderland		UKE3	South Yorkshire
	NL23	Flevoland		UKE4	West Yorkshire
	NL31	Utrecht		UKF1	Derbyshire & Nottingham
	NL32	Noord-Holland		UKF2	Leicestershire
	NL33	Zuid-Holland		UKF3	Lincolnshire
	NL34	Zeeland		UKG1	Herefordshire
	NL41	Noord-Brabant		UKG2	Shropshire & Staffordsh.
	NL42	Limburg (NL)		UKG3	West Midlands
Norway	NO00	Norway		UKH1	East Anglia
Poland	PL00	Poland		UKH2	Bedfordshire & Hertford.
Portugal	PT11/PT12/PT14/PT15	Portugal except Lisbon		UKH3	Essex
	PT13	Lisbon Region		UKI1/UKI2	London Region
Romania	RO00	Romania		UKJ1	Berkshire
Slovakia	SK00	Slovakia		UKJ2	Surrey
Slovenija	SI00	Slovenija		UKJ3	Hampshire
Spain	ES11/ES12/ES13	Galicia/Asturias		UKJ4	Kent
	ES21	Pais Vasco		UKK1	Gloucestershire
	ES22/ES23/ES24	Aragon/La Rioja/Navarra		UKK2	Dorset & Somerset
	ES30	Comunidad de Madrid		UKK3	Cornwall
	ES41	Castilla y León		UKK4	Devon
	ES42	Castilla-la Mancha		UKL1	West Wales
	ES43	Extremadura		UKL2	East Wales
	ES51	Cataluña		UKM1	North Eastern Scotland
	ES52	Comunidad Valenciana		UKM2	Eastern Scotland
	ES61	Andalucia		UKM3	South Western Scotland
	ES62	Región de Murcia		UKM4	Highlands and Islands
Sweden	SE01	Stockholm		UKN0	Northern Ireland
	SE02	Östra Mellansverige			
	SE04	Sydsverige	Not included	ES53	Baleares
	SE06	Norra Mellansverige		ES70	Canares
	SE07	Mellersta Norrland		FI20	Aland
	SE08	Övre Norrland		FR83	Corse
	SE09	Småland med öarna		PT20	Acores
	SE0A	Västsverige		PT30	Madeira
Switzerland	CH00	Switzerland			

Appendix B: Assignment of Patent Classes to the High-Technology Sector at the Four-Digit ISIC-Level

ISIC category	Industry sector	IPC patent category
3522	Pharmaceuticals	A61J, A61K, C07B, C07C, C07D, C07F, C07G, C07H, C07J, C07K, C12N, C12P, C12S
3825	Computers and office equipment	B41J, B41L, G06C, G06E, G06F, G06G, G06J, G06K, G06M G11B, G11C
3832	Electronics – telecommunications	G08C, G09B, H01C, H01L, H01P, H01Q, H03B, H03C, H03D, H03F, H03G, H03H, H03J, H03K, H03L, H04A, H04B, H04G, H04H, H04J, H04K, H04L, H04M, H04N, H04Q, H04R, H04S, H05K
3845	Aerospace	B64B, B64C, B64D, B64F, B64G

Appendix C: The Top-25 European High-Technology Regions: Patents, Citations Received and Citations Made (1985-2002)

	Patents		Citations received		Citations made	
	Number	Rank	Number	Rank	Number	Rank
Île-de-France	15,365	1	6,698	1	7,048	1
Oberbayern	11,278	2	3,344	4	3,277	7
Switzerland	7,486	3	4,956	2	4,856	2
Noord-Brabant	7,433	4	1,739	16	1,728	15
Darmstadt	5,878	5	4,438	3	3,948	5
Düsseldorf	4,787	6	3,164	5	4,322	3
Lombardia	4,726	7	2,873	8	2,540	8
Köln	4,529	8	2,903	7	3,986	4
Rheinhessen-Pfalz	4,013	9	2,513	9	2,447	9
Stockholm	3,802	10	1,771	15	1,885	12
Rhone-Alpes	3,635	11	1,492	20	1,401	19
Denmark	3,226	12	2,148	11	1,926	11
Stuttgart	3,184	13	1,102	23	1,130	23
Uusimaa	3,045	14	1,819	14	1,603	16
Karlsruhe	2,977	15	2,101	12	2,171	10
East Anglia	2,943	16	2,073	13	1,787	14
Berlin	2,754	17	1,262	22	1,264	21
London Region	2,531	18	1,582	19	1,420	18
Freiburg	2,424	19	1,713	17	1,504	17
Mittelfranken	2,192	20	1,375	21	1.031	24
Bedfordshire	2,178	21	3,107	6	3,375	6
Surrey	2,110	22	2,199	10	1,351	20
Schleswig-Holstein	2,034	23	748	24	1,213	22
Provence-Cote D´Azur	1,723	24	684	25	897	25
Essex	1,706	25	1,637	18	1,879	13

Figures

11   Patents, Patent Citations and the Geography of Knowledge
Spillovers in Europe

Tables

PART II   Innovation and Network Activities

5   The New Economy and Networking

**6   The Innovation Process and Network Activities of Manufacturing
      Firms**

**7   Knowledge Interactions between Universities and Industry in
      Austria: Sectoral Patterns and Determinants**

Subject Index

Author Index

Acknowledgements

The author and publisher wish to thank the following who have kindly given permission for the use of copyright material

Chapter 2
Reprinted from Alderman N. and Fischer M.M. (1992): Innovation and technological change: An Austrian-British comparison, *Environment and Planning A* 24, 1992, pp. 273-288, Copyright © 1992 Pion Limited, London

Chapter 3
Reprinted from Suarez-Villa L. and Fischer M.M. (1995): Technology, organization, and export-driven R&D in Austria's electronics industry, *Regional Studies* 29(1), 1995, pp. 19-42, Copyright © 1995 Taylor & Francis (http://www.tandf.co.uk)

Chapter 4
Reprinted from Cukrowski J. and Fischer M.M. (2002); Information-processing, technological progress and retail markets dynamics, *Information Economics and Policy* 14(1), pp. 1-20, Copyright © 2002 Elsevier, with permission from Elsevier Science Ltd.

Chapter 5
Reprinted from Fischer M.M. (2003): The new economy and networking, in D.C. Jones (ed.): *New Economy Handbook*, Academic Press, pp. 343-367, Copyright © 2003 Elsevier, with permission from Elsevier Science Ltd.

Chapter 6
Reprinted from Fischer M.M. (1999): The innovation process and network activities of manufacturing firms, in M.M. Fischer, L. Suarez-Villa and M. Steiner (eds.): *Innovation, Networks and Localities*, Springer, Berlin, Heidelberg, New York, pp. 11-27, Copyright © 1999 Springer-Verlag, with permission from Springer-Verlag.

Chapter 7
Reprinted from Schartinger D., Rammer C., Fischer M.M. and Fröhlich J. (2002): Knowledge interactions between universities and industry in Austria: Sectoral patterns and determinants, *Research Policy* 31(3), pp. 303-328, Copyright © 2002 Elsevier, with permission from Elsevier Science Ltd.

Chapter 8
Reprinted from Fischer M.M. (2001): Innovation, knowledge creation and systems of innovation, *Annals of Regional Science* 35(2), pp. 199-216, Copyright © 2001 Springer-Verlag, with permission from Springer-Verlag

Chapter 9
Reprinted from Fischer M.M., Fröhlich J., Gassler H. and Varga A. (2001): The role of space in the creation of technological knowledge in Austria: An exploratory spatial data analysis, in M.M. Fischer and J. Fröhlich (eds.): *Knowledge, Complexity and Innovation Systems*, Springer, Berlin, Heidelberg, New York, pp. 124-145, Copyright © 2001 Springer-Verlag, with permission from Springer-Verlag

Chapter 10
Reprinted from Fischer M.M. and Varga A. (2003): Spatial knowledge spillovers and university research: Evidence from Austria, *Annals of Regional Science* 37(2), pp. 303-322, Copyright © 2001 Springer-Verlag, with permission from Springer-Verlag

Chapter 11
Fischer M.M, Scherngell T. and Jansenberger E. (2005): Patents, patent citations and the geography of knowledge spillovers in Europe, in Markowski T. (ed.): *Regional Scientists' Tribute to Professor Ryszard Domanski, Studia Regionalia Vol. 15*, Polish Academy of Sciences, Committee for Space Economy and Regional Planning, Warsaw, pp. 57-75, Copyright © 2005 Committee for Space Economy and Regional Planning, Polish Academy of Sciences

Printing: Krips bv, Meppel
Binding: Stürtz, Würzburg